A Life's Design

The Life and Work of Industrial Designer
Charles Harrison

Foreword by Victor Margolin

Ibis Design Incorporated

Ibis Design Incorporated
6721-3 S. Merrill Avenue
Chicago, Illinois 60649

Copyright © 2005 by Ibis Design Incorporated.
All rights reserved.

Except as permitted under the United States
Copyright Act of 1976, no part of this publication
may be reproduced or distributed in any form or by
any means, or stored in a data or retrieval system,
without the prior written permission of the publisher.

Designed by Joeffrey Trimmingham.

Library of Congress Cataloging-in-Publication Data

ISBN 0-9773271-0-8

Printed in China

This book is dedicated to my parents for being patient with me and guiding me; to my wife Janet, who supported me through most of my adult life; and to my son Charley for his love, support, insights and help in pursuing this project.

Special thanks to Joeffrey Trimmingham for conceiving of this project and spearheading its development, and Victoria Matranga for encouragement and assistance through the rough spots.

I'd also like to extend a note of gratitude to my many friends both those within the design profession and those who know me personally and have helped make life meaningful.

TABLE OF CONTENTS

FOREWORD	i
DEFINING GOOD DESIGN	1
MAKING LIFE FROM SCRATCH	15
COMMITTING TO DESIGN	27
STARTING MY DESIGN CAREER	41
WORKING AROUND THE WORLD	71
GIVING OF MY GIFT	117

FOREWORD

When Chuck Harrison joined the industrial design staff of Sears Roebuck in 1961, he became the company's first black executive at headquarters, not to mention one of a small number of black executives in all of corporate America. Besides that, he was one of very few black designers in a profession that was notorious for the absence of people of color. Before Chuck, there were a few other black designers in large corporations. As examples, Georg Olden was the director of on-air television graphics for CBS in the 1940s, while General Motors hired McKinley Thompson as an automobile designer in 1956.

The paucity of black designers in America's postwar years is understandable in terms of racial attitudes at the time but it makes no sense when we consider that in the early years of the United States, black artisans, as art historian James A. Porter tells us, "became the backbone of American industrial development in the South and in parts of the Middle East [America]." Recent histories of American invention also show us that more black inventors than anyone ever imagined have contributed to America's manufacturing culture.

It is therefore no surprise that Chuck Harrison's design skills flourished during his thirty-three years at Sears. Some designers are known for the creation of two or three chairs or even a single chair, while Chuck worked on several thousand products that ranged widely from baby cribs and barber chairs to hearing aids, television sets, and tractors. As the first black designer at Sears and one of the few in American industry, Chuck demonstrated clearly that race is not an issue when it comes to talent and that, given the opportunity, a person of color can have an exemplary design career.

Before Chuck joined the Sears staff, he was fortunate to receive support from several mentors and employers: Henry Glass, Joe Palma, Ed Klein, and Bob Podall. While working in Podall's office, he redesigned the View Master, which still remains one of America's most popular icons. Chuck also benefited from the open atmosphere at the School of the Art Institute of Chicago, where a number of black artists and designers, including Charles Dawson, Archibald Motley, and Herbert Temple, studied both before and after he was a student there. And in Chicago there were other black designers who made their mark on white corporate culture: LeRoy Winbush, whose company Winbush Associates designed almost all the display windows for the city's banks, Tom Miller, a talented graphic designer who worked for Morton Goldscholl Associates, and Andre Richardson King, who started the signing department at the architectural firm, Skidmore, Owings, and Merrill.

In this book, Chuck Harrison tells the inspiring story of his life. The product of a strong family and community, he was able to pursue his goal to become a designer despite all the familiar obstacles that faced a black person in the fifties. As a design professional, Chuck is a true pioneer, not only because he broke the color barrier at a large American corporation but because he designed so many of the products that became and still remain part of the daily lives of millions of people. As the history of design continues to include more designers beyond the familiar icons, Chuck will surely have a place as a leader in the design of mass-market consumer goods. He has taken the first step by telling his story. What should follow is the wider recognition of all the amazing things he did in his long and illustrious career.

Victor Margolin is Professor of Design History at the University of Illinois, Chicago.
© 2005 Victor Margolin

"""The first word that comes to mind in describing Chuck is integrity. He's completely honest, not only in his personal relationships but in approaching a job. That's what made him an outstanding designer: no short cuts, no bull, no fluff that doesn't mean anything. He always fought against that in a marketing environment that frequently put him in conflict. Chuck still thinks a lot of what we deal with is less than real, less than meaningful. He's actually a pretty simple guy."

~Bob Johnson was a successful Sears & Roebuck senior vice president. He co-founded Johnson Bryce Corporation in 1991.

DEFINING GOOD DESIGN

AN EYE FOR EVERYDAY ESTHETICS

A few years ago, I attended the opening of the new public library in Evanston, Illinois, a suburb of Chicago. The event featured the unveiling of a commissioned sculpture by internationally renowned artist Richard Hunt, who was one of my fellow students at The School of Art Institute of Chicago during the early 1950s.

As I stood near the refreshments waiting to speak with Richard after the ceremony, I spotted something remarkable. The director of the Center was in earshot, so I said to her, "If I had known you guys were going to display my artwork here, I would have autographed it."

"You have a piece of work here?" she asked.

"Yes," I said. "Right over there—that garbage can."

"Oh, is that your can?" she replied. "If we had known that, we would have cleaned it up!"

As it turns out, she had attended art school in Detroit and knew about industrial design.

I designed the can in the mid-1960s, while working for Sears, Roebuck and Company. Looking back at my career, it was one of the most significant and innovative products I ever created. When that can hit the market, it did so with the biggest bang you never heard—everyone was using it, but few people paid close attention to it.

ABOVE: Prior to 1966, garbage cans were made of metal. For the most part, they were either old 55-gallon drums or 20- or 30-gallon round galvanized cans. Not only did they rust, but, on garbage pickup day, they made a noisy racket in neighborhoods all over America! Also, as they were hit by cars and banged around, they looked worse and worse.

This first of its kind, plastic garbage container was designed for Sears in 1966 and has been used by more people world-wide than any other product I designed. Because it was necessary to ship large numbers of containers, I designed them so they could nest inside each other. If the 30-gallon cans were shipped separately, 20 or 30 of them would probably fill up an entire trailer truck, but since they nested together, the same truck could carry several hundred.

(and because it's all heavy-weight plastic, there was no noisy metallic clang).

Yes, Sears has everything
..and everything is so easy to buy when you have a Sears Credit Card

DESIGNING THE CAN

In the early 1960s blow molding was increasing as a process for making housewares, and Sears was often at the forefront of product innovation and development. Richard Palase, a chemist in Sears' laboratory, proposed making a refuse container of blow-molded polypropylene, and recommended it to Alan Karch, a Sears buyer. With Karch's support, Sears assigned a team of technical professionals to the challenge. I was the designer on the team, and after numerous planning sessions with the prospective manufacturer, we established the product direction. As with every product, though, the devil was really in the details. We needed to meet the recommended criteria: a dark color was recommended to withstand the outside weather and ultraviolet rays; the container needed handles for transporting from one location to another; a sloped lid shape to allow rain or melting snow to run off; and hand grips on the bottom for ease when emptying.

The container would also have to withstand heavy impact from dropping and hitting. The lid would be designed to make it easy for the owner to remove, but difficult for animals such as dogs or raccoons. The shape needed to be one that nested, in order to maximize shipping quantities. And, the surface had to be textured to assist in preventing scratches from shipping and usage.

To convince people how much better this can was than previous ones, we put together a test: We froze the can at 40 degrees below zero for a couple of days, put a 50-pound bag of sand in it and threw it off the top of a five-story building. The thing didn't break! It just bounced.

We knew we were onto something, so Sears marketing decided to coordinate an even bigger stunt: they dropped the can out of a helicopter, and it performed well again.

The refuse container was granted a patent for design details of its lid, which resisted opening if the container fell or was knocked on its side. The product was very successful and profitable for Sears.

Other design generations followed which added wheels and a rectangular shape to accommodate better use of interior space with filled grocery bags.

OPPOSITE: In 1966 Sears ran this ad illustrating the durability of our refuse can. The can had been frozen for several days and was dropped from a helicopter to see how it would hold up. It passed the test with flying colors. More than 30 years later, I chanced upon a can from that original line.

ABOVE: I designed a rectangular version of the can to accommodate the paper grocery bags that were commonly used as garbage bags in mid-1960s American households.

DEFINING GOOD DESIGN

The Sears catalog detailed some of the main attributes of the refuse can.

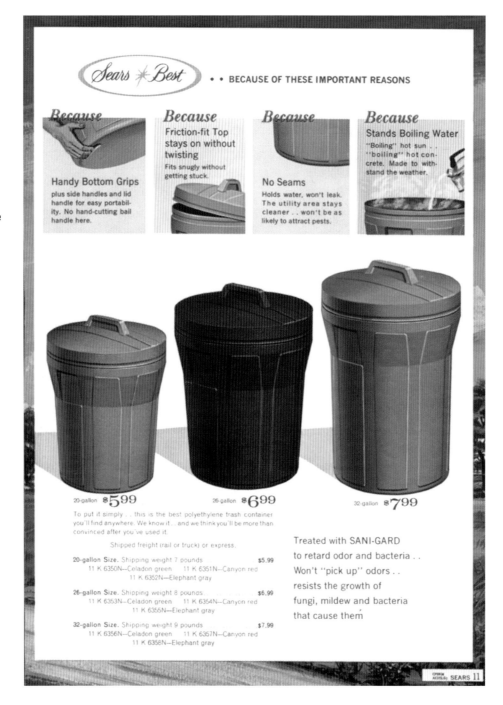

6

Throughout my career as an industrial designer, which included 33 years at Sears, 85 percent of my effort went into designing consumer products to improve people's everyday quality of life.

I designed everything from binoculars to baby cribs, televisions to toothbrushes and almost everything in between, including a lot of sewing machines. In fact, someone who had heard about my work once said, "You designed all those sewing machines? Well, you must be the Michael Jordan of sewing machine design!"

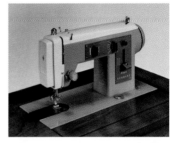

From sand-casted iron to injection-molded plastic, I designed 8–12 sewing machines every year, for about 12 years.

DEFINING GOOD DESIGN

OPPOSITE AND THIS PAGE: My early work focused on furniture design, which played a strong role throughout my career. Among the many sewing machine cabinets I designed were the desk- and side-table models shown.

The two Sears Catalog pages shown promoted the features and uses of several of the sewing machines I designed.

DEFINING GOOD DESIGN

Sell sheets like this one shared many of the features of this versatile, portable, convertable sewing machine.

New from Sears! A sewing machine that's convertible.

Flat surface for all your regular sewing.

You also get all these features:

A snap-in buttonholer that sews 5 different kinds — including keyhole. No other machine sews as many — so easily.

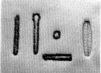

You get 44 different stitches — including 7 stretch stitches.

Solid-state controls give you full power, even at the slowest speeds.

With the no-skip needle and foot, you may never skip a stitch again — no matter how slick the fabric.

There are at least 20 other features you ought to know about.

Free arm for the things you used to sew by hand.

Sleeves, armholes, cuffs and collars slip right on. So do pants legs — without ripping the seam. Perfect for blind hems, too.

Shop around. Compare machines. You'll soon see what a *value* Sears offers. And don't forget Sears service and dependability. Get a demonstration at most Sears, Roebuck and Co. stores.

Also available through the catalog: #1914.

| Sears | **Kenmore Sewing Machine**

This serger was designed for the Maruzen Company in Japan. The sewing machine provided special stitching capability beyond that of most basic sewing heads.

Jaguar brand products were designed for export sales to the USA and Europe. (1993)

DEFINING GOOD DESIGN

Arguably Michael changed the game of basketball, but I don't think designers can change the world. Rather, they can take what's here and make the best of it. Form and shape work best when they just seem to be there, not forced. I tried to make things appear as if they just belong; that they didn't need to scream, "Look, here I am." My best efforts resulted in products that did their job as expected—you look at it, right away guess what it is supposed to do, and that's exactly what it does. Or maybe I should say that the visual statements express a harmony with why the product exists—what it does; how it is made; what it is made of—while looking pleasing if not beautiful.

So much of life factors into the design equation: business considerations, the social and natural sciences, art, engineering or communications. Designing for me is living with an understanding and sensitivity to these areas, and having the ability to solve a specific need—like the need for a quieter yet durable garbage can that didn't rattle and bang loud enough to wake the dead on a quiet suburban street of the 1960s.

I once had a plaque over my desk that read: "How do you define a designer? You don't. You describe him."

The plaque went on to describe a person with an eye for esthetics and a concern for profits; who understood production and cost problems; and who had a complete working knowledge of many materials. Above it all, it described a person who prefers to design from the inside out because he was every bit as concerned with the product's function as he was with the product's appearance. This designer believed that esthetic appeal is based on cold-hearted practical considerations.

And rather than define who I am, I'd like to once again rely on description to shed pale light on how, despite a long list of despites—despite being black, despite being poor from the rural South, despite being dyslexic, despite being born in an America still caught up in its own growing pains—I came into my own as an artist and human being, dedicated to giving back through my craft, and through the act of teaching that craft to others eager to carry on the legacy of good industrial design.

MY LIFE'S WORK
Over the course of my career, I've been involved with the creation of almost all areas of household products.
Below is just a short listing of the many products I've designed.

- 35mm cameras
- AM-FM radios
- baby cribs
- back massagers
- barber chairs
- bicycles
- binoculars
- blenders
- calculators
- can openers
- cassette recorders
- circulating heaters
- clothes hampers
- coffee percolators
- compact whirlpools
- cordless shavers
- dinette sets
- electric mixers
- electric scissors
- electric toothbrushes
- electric wall clocks
- fishing equipment
- fondue pots
- hearing aids
- hedge clippers
- insoles for shoes
- kitchen appliances
- kitchen ranges
- makeup mirrors
- manicure tools
- paint brushes
- phonographs
- portable bars
- portable hair dryers
- riding lawn mowers
- shoe buffers
- sleds
- soda fountains
- sprinklers
- storm doors
- telephones
- television sets
- toasters
- tractors
- waffle irons
- wall unit shelving
- wet mops
- window guards
- wooden desks

In the 1960s, I photographed this boat near Venice, Italy, while on an exploratory Sears trip to find a manufacturer for wall systems. At the time, only one American company had the capacity to manufacture ready-to-assemble wall systems. During my career, my buyers and I traveled all over the world to find manufacturers capable of delivering for Sears.

"Charles was a pretty smart kid, pretty innovative. He'd tell a lot of jokes, as did his Dad. Charles was just nice to be around, very friendly, got along with everybody. I especially remember that we were both interested in building model airplanes, and I learned a lot from him. He always made his airplanes very neat. He used just the right amount of glue. Things weren't smeared all over. For a child, the Prairie View campus was an ideal world. As you could expect, there was a lot of segregation and prejudice in the time we grew up, but being isolated, we were sheltered from a lot of that, at least until we would leave the campus—not that we weren't aware of it."

~Ernest Norris, a childhood playmate, lives in Prairie View after a career as the production manager of a weekly newspaper and an advertising staff member of a department store

MAKING LIFE FROM SCRATCH

BATON ROUGE, LOUISIANA

I was born in Shreveport, Louisiana, in 1931, in my grandparents' home. Most African-American children, in those days, were born at home. At the time, my parents—Charles Alfred Harrison Sr. and Cora Lee Smith—lived just outside Baton Rouge, where my father taught industrial arts at Southern University. I had one older brother, Lawrence. Even though he was nine years older, we shared a lot together.

My family visited Shreveport in the summers and sometimes on holidays. It was the second-largest city in Louisiana, but was considerably different from New Orleans, which had a French character and a party-like lifestyle. Life in Shreveport was fundamental. Most of the people were illiterate. In fact, the buses and trolleys had symbols on them—squares, circles, and so on, so people would know which one to take because they couldn't read the written signs.

My grandparents lived in a very poor part of town, as almost all the black people did. People didn't go to the hospital unless they were dying, but instead treated themselves at home. Once, a nail went through my foot, and my folks just wrapped my foot in some kind of flour sack and soaked it with turpentine. We took baths on Saturdays in a tin washtub; obviously, the last guy didn't have very clean water. We had very few store-bought toys. We made skate boxes—equivalent of today's skateboards—with an old piece of two-by-four and old skate wheels. That was a big deal.

I don't remember much of life near Baton Rouge. It was a very small community, and we didn't have all that much. It was during the Depression, and my father earned $40 a month for the four of us. But we had food and clothes, and the university provided housing for faculty. Our house was right on the small campus.

PRAIRIE VIEW, TEXAS

Shortly after I turned five, we moved to Prairie View, Texas, home of Prairie View A&M University. My father worked as a professor at the university, which was located about 45 miles west of Houston on a line between Houston and Austin. The nearest town, Hempsted, was six miles away. It had one signal light, a movie theater, a grocery and a barbershop. But when we really wanted to go to town, we went to Houston.

In Prairie View, I began to flourish. It was rich in things to do. The rural setting accounted for few fences which gave us lots of room to roam. Since there weren't a lot of children—only the kids of the faculty and staff and a few farm children from the area—I was a child of the neighborhood. Anyone around there had the authority to discipline us, but they also all cared for us. We were closely observed—although we did manage to find a few moments to have fun.

MAKING LIFE FROM SCRATCH

My whole family in our house at Southern University (circa 1935).

My 6th birthday party in Prairie View.

Riding my horse on the Prairie View campus.

Originally, Prairie View had been what was called a "Normal" college, set up to educate students who were going into teaching. So-called Normal schools were established to train students to become teachers. By the time we arrived, Prairie View was an agricultural and mechanical college—hence A&M. Although it continued to produce a lot of teachers, it also graduated more African-American engineers than any other school in the country.

At Prairie View, we were all African Americans, so there was no segregation. We saw white people only when they came to the hospital or brought us deliveries.

The place was very self-sufficient. It had its own dairy, hospital, veterinary hospital, bakeries, pharmacist, service station, fire department and even a one-man police department.

The school also had ROTC units, which trained military officers. (There were still segregated units.) Sergeant Harris, a soft-spoken man who was an instructor, lived two doors from us and had been in the Buffalo Soldiers, a cavalry unit in Arizona. He taught me how to ride and care for horses and took me to the rifle range to teach me how to fire weapons. He was a friend of the family, and had no children of his own. I liked this soft-spoken man. He fueled my fascination with horses and guns, and I was eager to learn military things.

At a very early age, I started getting little jobs around the college, like working for painters and for the laundry. I also had the opportunity to go into the veterinary hospital and watch procedures on animals and go into the poultry yards to learn about chickens.

I even herded cattle on the horse my father bought me. We had all kinds of animals at home like pigs and

On a float with other Prairie View children in a Homecoming parade.

With my parents on the front lawn of our home in Prairie View.

chickens. We sold eggs from our chickens, and I delivered them. I rode on the milk "truck"—a milk wagon pulled by a horse—to help deliver milk.

I remember one day a guy flew an airplane and landed in Prairie View, even though there was no airport. He landed right on a little road and then had to move his airplane so cars could get by. We were all "oo-ing" and "ahh-ing." I was fascinated by airplanes and built models of them, so my father asked the pilot if he could take me up. We went up and flew one circle around the campus and landed in a field. I must have smiled for a week after that.

Even through Prairie View was a very small community, the college campus provided all kinds of entertainment, including plays, concerts, recitals, athletic events and statewide Boy Scout Jamborees.

I saw some outstanding tennis players, and played a lot of tennis myself. I started at age eight, using an undersized racket, and later played in high school, college and the Army. I had a pretty good game, and continued until the mid-1990s. I also saw people like Marian Anderson and Paul Robeson come in for concerts. I was just a little squirt, and they'd play with me—pick me up and talk to me.

Growing up, I admired my father and grandfather. My father, in addition to teaching, was sort of the general contractor for the campus. He was fundamentally a carpenter, but he also knew electrical and plumbing systems. My mother's father was also a carpenter, and I often trailed behind him as he put this big toolbox on his back and walked everywhere to do his work. So my ability to work with my hands came from both my grandfather and father.

I felt a sense of accomplishment when I worked with them, seeing projects completed and hearing the compliments from people whose lives we made easier and better. In later years as a designer, this sense of fulfillment grew stronger as my work expanded from one-to-one contributions to improving the welfare and well-being of thousands of people through mass-produced products. This is a fundamental part of who I am. I honestly believe that I'm on this earth to help others and this work was my purpose.

My father also built all of the furniture in our house—tables, beds…everything. He built wagons and gym sets for us, and once even made a harness for our goat to pull a wagon.

My mother was always busy. She cooked our food, washed our clothes, and managed the family store that sold chickens and eggs. My mother also belonged to a bridge club and hosted many parties. At our parties, my brother played piano, and I did a few flips for the guests.

PHOENIX: LEARNING TO SURVIVE

We lived in Prairie View until 1945, when my dad got a job teaching at an all-black high school in the Phoenix school system.

The school had opened in the early 1920s as Phoenix Union Colored High School. In 1944, the school system changed the name to George Washington Carver High School and hired a lot of new teachers and a new principal from Atlanta, Professor Robinson. He hired mostly college-level teachers.

Eventually, every member of the faculty, including the physical education teachers and the shop teacher, had master's degrees. And those teachers really took us seriously. They were dedicated to doing things for us and with us. It showed what the reverse side of segregation could do for you. It didn't compensate for the negatives, but it was something positive.

Carver closed in 1954, when integration became law. It was then used as a book depository, until the mid-1990s, when an alumni group bought the building from the city. Now it houses the George Washington Carver Museum and Cultural Center, which pays tribute to contributions made by African Americans from Arizona. One room has been dedicated to my work!

Moving to Phoenix was a culture shock and my introduction to urban life. I was in the city, even though at the time Phoenix was a hick town. Unlike Prairie View, I had to stop at corners and wait for the stoplight.

Looking back, Arizona in the 1940s may have been even more racist than Mississippi. We lived in a duplex, and our neighbors on both sides were Mexican. But they had to attend a separate school from us. It was jokingly said that most blacks who lived there had gotten off the train on the way to California and hadn't gotten back on by the time the train left.

Since our school was the only black high school in the state, so students had to travel to Phoenix just to attend high school. Two of my friends, Corey and Ruth Payne, traveled from Chandler, which was 50 or 60 miles away from Phoenix.

They would get on the bus from home and get off at the bus station in Phoenix, walk another two miles to school and at night do the whole thing in reverse to go back to Chandler, every day for four years.

After graduating from college, Corey got a job teaching in Chandler. After two years, he became a member of the Chandler board of education. And several years later, he became mayor of the city where he wasn't allowed to attend school!

I was very involved in high school: the band, chorus, basketball and the tennis team. We competed against white schools because there was no one else, but we rarely had a problem. At that time, Phoenix also had separate schools for Native American students, Asians, Mexicans, whites and affluent whites. Ironically, summer school was integrated. We attended summer school with students from other minority groups. Apparently, the city fathers didn't want to pay for separate summer schools.

At the churches, we had socials together and actually danced with each other. The only other time we could interact was in the parks. Two or three white kids couldn't play a game of baseball by themselves, so they and the blacks and the Mexicans would join up. And it was fine.

Even though our high school had only about 250 students, we won the city basketball championship in 1948, which meant we got to compete in the state championship in Tucson at the University of Arizona. But ours was such a poor school that we didn't even have basketball uniforms! Our coach told our principal, "I just can't write a number on their T-shirts like I did during the season." The principal told him to go to the sporting goods store and see if they'd sell uniforms on credit.

We went to Tucson representing the largest city in the state with the wrong-color uniforms, but we were proud to have them on. Also, because we were able to get only eight uniforms, when the time came to substitute, the coach had to send us to the locker room to change clothes. He would then tell the scorekeeper that the No. 6 going into the game was not the same No. 6 as before, that the new No. 6 had no fouls on him. This went on until, after awhile, the scorekeeper said, "Hell, you don't have to tell us anymore. We know who they are." As I remember, we won a game or two, but didn't go all the way in the tournament.

Anyway, when our high school students graduated, they could go on to college. Ironically, going to college in Phoenix was not a problem. There were no restrictions or segregation while attending college.

I mean, the whole segregation business was just idiotic. It's interesting to note the contrasting ways Arizona treated the soldiers of our wartime enemies. Even German prisoners were given privileges denied to all of the minorities.

This photo of me was taken shortly before I graduated from high school in 1948.

MAKING LIFE FROM SCRATCH

High school graduating class from Carver High, Phoenix, Arizona, the only black high school in the state; I graduated from high school at age 16. (1948)

For instance, the city even separated the times we could go to the circus! They had a day for the blacks and the Mexicans and the Native Americans. If the Ringling Bros. Circus was in town for a week, we blacks could only go on one day. German prisoners went on any day they wanted. I have a Japanese friend, George Kimura, who was placed in an internment camp near Phoenix with his family during World War II. Although they were citizens, they couldn't even leave the camp much less go to the circus.

As a whole, life in Phoenix was difficult. It was really a tough life, and my father didn't like it. He wasn't paid much for teaching, and it was a different lifestyle than he wanted. It was like living in a ghetto; we were just poor and everybody had to hustle. My mother got a job working in Barry Goldwater's department store. I delivered papers and painted houses. My father worked in a nightclub at night as a cook and taught during the day. After about three years, my parents moved back to Texas. Since I was in my senior year of high school, I stayed to finish.

SAN FRANCISCO

When my parents moved back to Texas in 1947, they left me in Phoenix to finish my senior year. I stayed at the home of an elementary schoolteacher. About that time, my brother, Lawrence, returned from the war and started graduate school at the University of California at Berkeley. He convinced our parents that I shouldn't go to an all-black college, so, when I graduated from high school at age 16, I moved to California to attend college.

In high school, my life centered on day-to-day survival, so I had no career goals when I graduated. I just went to college. I thought I would get into Berkeley, but I didn't score well. I had a lot of difficulties—and still do—reading and learning. My fundamental problem wasn't discovered until much later: I was dyslexic. We didn't know about dyslexia back then, but my parents knew something was wrong, and tried to solve it by putting eyeglasses on me when I was about nine years old.

Since I didn't pass the college entrance test, Lawrence arranged for me to go across the bay to The City College of San Francisco (CCSF). I lived in Oakland with Lawrence and his wife and took a train to San Francisco every day, then took the streetcar to the campus. It was a two-year college, but more than 80 percent of students went on to Stanford and UC Berkeley. The students at CCSF were sharp and there was another shock: the school was almost totally white. There were probably 20 black students out of about 1,000. This was my first experience of feeling completely isolated.

After an unsuccessful semester as an Economics major, I needed some direction and found it in a course called Vocational Guidance. The course was designed to help students find careers.

With my sister-in-law, Ora, in Oakland, California, when I first started college; Ora Scott was married to my brother and now lives in Austin, Texas, where she retired from the state board of education. (1948)

MAKING LIFE FROM SCRATCH

At the end of the semester, the vocational guidance instructor evaluated my scores, and he thought I would do well in art. At that point, I needed a life preserver, so the next semester I registered as an art major.

Much to my surprise, things really started to turn around. When I got a "B" at the end of my third semester, it was time to celebrate. My grades continued to improve from there, and by the time I completed studies my second year at CCSF, most of my grades were A's. By then, I was 18, and had a growth spurt, which also helped a lot with my confidence. I not only had been 16 when I started college, but I was smaller and didn't fit in with the other students too well. The girls were not only older than m, they were taller.

As the end of my time at CCSF got closer, the head of the art department asked me what I was going to do to continue my education, and I said I was either going to UCLA to study interior design or go somewhere else for industrial design. He said he thought I'd be a good painter. I said, "Thank you, I appreciate the compliment, but if I went home and told my father I was going to be a painter, he'd say, 'Well, son, that's wonderful. You can get started on the garage right now.'"

After leaving CCSF, I went back to Texas and started looking into industrial design. At that time only five schools in the U.S. held accredited industrial design programs: Rhode Island School of Design, Pratt Institute, Carnegie Tech, the School of the Art Institute of Chicago and the University of Illinois at Urbana-Champaign. Since it was very difficult to get into colleges at that time because World War II veterans had priority, I applied to all five. I was accepted at the School of the Art Institute of Chicago (SAIC).

Around that time, Lawrence moved to Houston, where he taught psychology at Texas Southern University. He and my parents helped me get the funds together for college.

In many Southern states, if a black student wanted to study a field that was not offered at a black college, the state would pay his tuition and transportation to go somewhere that did offer it. As long as I didn't go to school in Texas, they didn't care. They probably would have sent me to Paris or Rome! That's how I was able to finance my undergraduate education. I entered the School of the Art Institute in 1949, on a four-year tuition and travel scholarship provided by the State of Texas.

My schoolmates and I enjoy a meal and some sun in Dallas, Texas, during Christmas break. (1956)

"Chuck was a terrific student, and did very worthwhile projects in school. He was meticulous in what he did and was persistent in learning. He also introduced me and other classmates into volunteering for a South Side home for disadvantaged children. We'd take these kids out to the lakefront a number of times."
~Dick Humbert, of Park Forest, Illinois, a retired industrial designer and classmate at the School of the Art Institute

Even though I was the only black student in my classes at the School of the Art Institute (SAIC), my first year proved to be different from what I expected. Unlike CCSF, my fellow classmates accepted me, and there was a genuine sharing of information as well as social activities. Somehow, in the arts, racism seemed to be less intense than other places. I guess the sharing of information didn't represent power in this field, so the guys in my class were very helpful. The first year I became a member of the Honor Society.

I did well at the School of the Art Institute of Chicago. I remember the day in class that I truly realized the strength of my skills. The guy next to me was struggling, maybe more than I was. He was an older guy, a veteran. Our instructor really chewed this guy out. I mean, he tore him apart. I said to myself, "I'm better than this guy. And he's white. I'm going to be OK." I realized that I really had a chance to compete.

My experience in that class, particularly with the instructor, Henry Glass, a strong force in U.S. furniture design, kindled my interest in furniture design. All of my future employers hired me on the strength of my furniture design skills.

During my sophomore year at the School of the Art Institute, I met a girl, Janet Simpson, at a party thrown by mutual friends who thought we two out-of-towners needed to meet. We met through two brothers, Willie and Eddie Milan who decided to give a party and used the occasion to acquaint Janet and me. Willie attended the SAIC with me and Eddie attended the Century College of Medical Stenography with Janet. The brothers were from Canton, Ohio and Janet had moved to Chicago from Indianapolis. She and I hit if off and we dated for a couple

ABOVE: For a school project in 1954, I designed and built this chair during the height of the modern furniture design movement. I selected curly and hard maple for the chair and used brass rods for the back to introduce another material for visual accent.

The design emphasizes purity and honesty in furniture through the use of materials in their natural state and finished with clear coatings to reveal the grain and true color of the wood. Also, good craftsmanship is stressed by exposed spline joints in the rear legs for additional interest and expressions of quality.

This model is now in the archives of the DuSable Museum in Chicago.

COMMITTING TO DESIGN

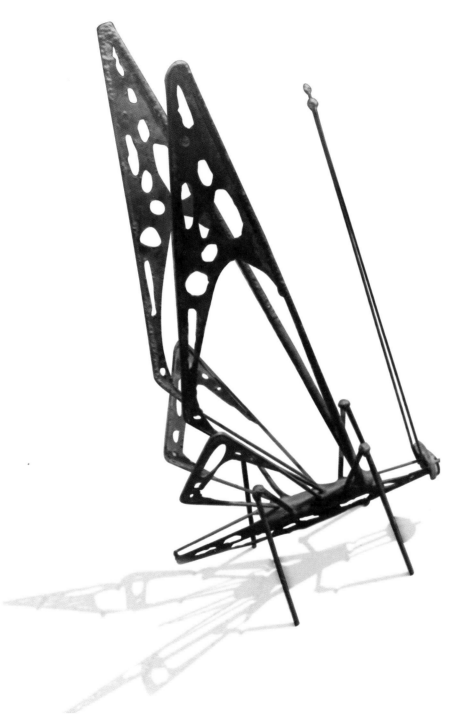

During my undergraduate years, in addition to developing an understanding of texture, color and other design elements, I had the opportunity to work with tools like welding torches, which I had learned to use in Prairie View. Doing sculpture was a real pleasure. This gave me the opportunity to explore an esthetic approach for three-dimensional work, creating fine art rather than functional objects. Some of my undergraduate projects included:

LEFT: "Butterfly," a welded steel sculpture that I sold to a New York writer (1953),

MIDDLE: A plaster sculpture of a giraffe based on a photo I found (1953), and

BOTTOM: A volume study in plaster (1953).

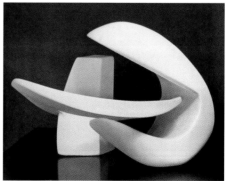

RIGHT: This globe was a student project submitted to the Rand McNally competition. It is now on display at the George Washington Carver Museum and Cultural Center in Phoenix, Arizona. The design included a brushed aluminum centerpiece with wrought iron legs. There was a mechanism in the hub that allowed the globe to turn. (circa 1954)

BELOW: This magazine rack was a student project during my undergraduate studies at the School of the Art Institute of Chicago. (circa 1954)

COMMITTING TO DESIGN

Dinner with friends and my future wife, Janet. (1952)

My induction into the national art honor fraternity, Delta Phi Delta, at the School of the Art Institute of Chicago. (1953)

of years but later drifted apart. I didn't know it at the time, but Janet would later be my wife.

During the final two years of my time at the SAIC I studied with Joe Palma, a designer recognized in the industry for the high level of esthetics in the products designed by his firm. Under Palma, I refined my skills and developed an exceptionally high appreciation for product design esthetics. And, at the end of the four years, I graduated with very strong grades.

It was at the School of the Art Institute that I acquired the knowledge and skills that really made me a designer. I learned the principles of design, esthetics, materials, manufacturing processes, concern for the client and the constant need to keep the end user in mind from professionals like Henry Glass and Joe Palma. They shared information about the field which couldn't usually be learned in the classroom. They allowed us to visit their offices and see works in progress, and even hired some students. I also developed a strong foundation in the fine arts, learning design, color, form and drawing.

This was when I first really knew that I wanted to be an industrial designer. I was able to relate all the skills I had, including the creative things that my father and grandfather had taught me. Once my father understood what I was doing, he was really excited. I don't know if I was surprised, but I was really relieved to have my parents' support. Oh, they would probably have supported me as a painter. But I didn't want to test them!

In December of 1953, I had satisfied all the requirements for graduation, but I thought I needed more credits. So I registered for the next quarter, during which time I was able to get a job designing interiors for Maurice Sternberg. Maurice had an office on Walton Street right next door to the first Playboy Club in Chicago. He wanted someone who could design furniture and I had those skills, which few interior designers did at that time.

I joined his staff for four or five months, and designed for residential homes in the northern suburbs. We also designed showrooms in the American Furniture Mart for manufacturers like Rowe and Broyhill. We even designed

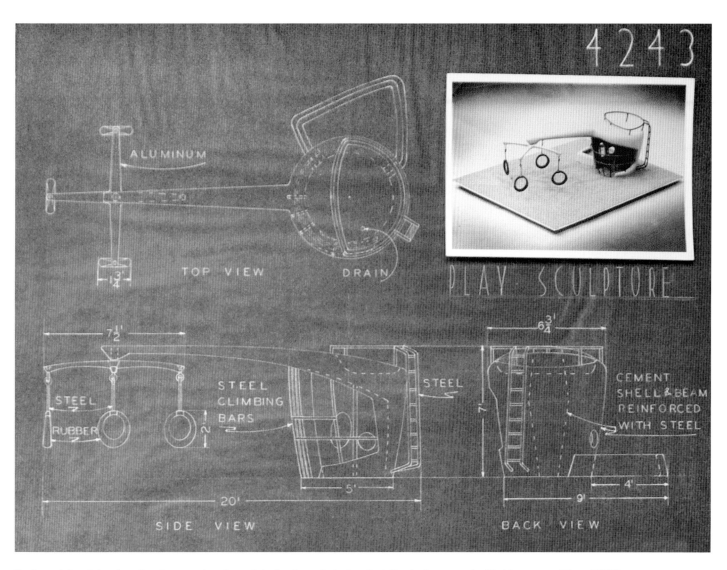

Scale model and drawing of a playground equipment design for a student project; the design was submitted to a competition. (1953)

COMMITTING TO DESIGN

model apartments for new high-rise apartment buildings, including 1000 North Lake Shore Drive on Chicago's Gold Coast. My primary responsibility at Sternberg's, however, was to design custom furniture for his clients, including bedroom suites, tables, and cabinets. I spent most of my time working on case goods.

In March 1954, my studies ended and I was drafted into the Army. I continued to work for Maurice until I graduated from school and I reported for duty in June.

OPPOSITE AND ABOVE: These arrangements were created as part of my work with Maurice Sternberg for the 1000 North Lake Shore Drive apartments. The designs appeared in model units and were featured in *Chicago Tribune* Magazine. (circa 1954)

COMMITTING TO DESIGN

With Dad, Mom, and Lawrence in Chicago on the day following my graduation from the School of the Art Institute of Chicago; my brother lived in Chicago and our parents had driven from Dallas. (1954)

Graduation ceremony from the Combat Engineers Basic Training at Fort Leonard Wood, Missouri. (1954)

COMMITTING TO DESIGN

MILITARY SERVICE

At Ft. Belvoir I learned to be a map maker, and by December I was in Frankfurt, Germany. I was assigned to a section of a topo (topography) unit that did spot mapping and drafting. We created military maps by combining aerial photographs with survey charts. Well, some areas that were photographed, particularly in Bavaria, were heavily wooded. It was impossible to identify what was below the trees. So they assigned a team of approximately 10 people, most of whom were surveyors, to go into the field and draw these portions of maps. I was the only draftsman.

The guys in my unit were sharp. They were all graduate engineers, and some had worked for state highway departments. After we finished mapping West Germany, our assignment was to identify targets for guided missiles and aircraft all along the East German and Czechoslovakian borders. The engineers recorded coordinates to pinpoint the location, and I drew detailed illustrations showing the targets and all the landmarks and structures around them. We saw a lot of bombed-out cities. So much of the country was just rubble.

As we did our work, we tried to find nearby army installations where we could sleep and eat. If they didn't have space, we'd have to sleep in the woods or in trucks, eat "C" rations and probably indulge in too much alcohol. We sometimes went into a village near where we were working, and take the cigarettes and chocolate out of our rations and trade them for beer, wine, wurst and bread.

I didn't take advantage of as many things as I could have in Germany. But I took lots of photographs, using an Army lab to process them. In fact, I entered competitions and won prizes for my photography. I also painted watercolors at night after striking up a friendship with a local professor who had once taught in Hamburg. He taught me

Drawing maps in the field, in West Germany. (1955)

Although my post wasn't a combat position, I sometimes served guard duty on Army posts and other installations. (1954)

his watercolor technique.

One day we were traveling in small trucks with two mechanics who had been assigned to us as drivers. One of them was kind of lax in the way he did things, and he drove us out on a strip between East and West Germany. Either the East Germans or Czechs opened fire on us with machine guns and shot holes in our truck. Fortunately, no one was hurt. But it was a close one.

I was in Germany for almost two years, and it was good duty, but I didn't want to be there. I started thinking about what was next for me, so I reestablished contact with people I knew. Around this time, I wrote to Janet, who had moved back to Indianapolis, where her parents lived. We started communicating frequently and rekindled our interest in each other.

One night in December of 1955, at an NCO club where troops gathered after work, I learned that service members could be released from duty as much as three months early with the condition that they would return to graduate school. I pursued schools in Ecuador, Brazil and Colombia with the goal of entering architecture school. But those schools' schedules did not match my timing requirements for early release from military service.

As a result of the combined desire to get out of the Army and the feeling that I had spent two years away from the work I really wanted to do, I wrote to Henry Glass, Joe Palma and the dean of the School of the Art Institute. The School didn't have a graduate program in industrial design, but Henry and Joe put one together for me.

As far as I know, there hasn't been a person before or since who studied graduate industrial design at the School. The schedule at the SAIC worked perfectly for me to meet the termination date for an early release from military duty, and in March 1956 I was honorably discharged from the Army three months early.

Playing the guitar during a moment of relaxation in Furth, West Germany. (1955)

The members of my unit gathered to celebrate news of my going home. (1956)

"Chuck was very straightforward, looking toward tomorrow. His designs were always very pleasing and creative. I would say that if he was not at the top of his class, he was pretty close to it."

~John Pryer, a schoolmate at the School of the Art Institute of Chicago, is a retired photographer now living in Menlo Park, California

STARTING MY DESIGN CAREER

In March 1956, I returned to the School of the Art Institute as a graduate student. My brother, Lawrence, lived in Chicago, where he was chief of the diagnostic clinic at Chicago State Hospital. He and his wife were separated and he was living in the YMCA. I stayed there with him for about a month, then we moved into an apartment in Hyde Park with Lloyd Keyes, a friend from CCSF. Lawrence worked during the day and took courses at the University of Chicago at night while I began graduate work at the School of the Art Institute.

Janet and Lawrence drove to Ft. Sheridan to meet me on the day I arrived to be discharged. I stayed at Ft. Sheridan for a couple of days to finish processing, then moved back to Chicago. Janet had already been living in Chicago for a short time, and we started seeing each other on a regular basis. We were married in August of 1957 and stayed married for more than 40 years.

As part of my graduate work, I did a project on electric ranges for Hotpoint, where Joe Palma had a connection with Ray Sandin, who was the manager of industrial design. Henry also assigned a very extensive furniture project to me. I designed a complete line of "knock-down" or "KD" furniture—today, they call it "RTA" ("ready to assemble")—including living room, dining room and bedroom pieces.

After a semester, my money ran out and I had to get a job. I looked around for a place where I could study at night and enrolled at the Illinois Institute of Technology

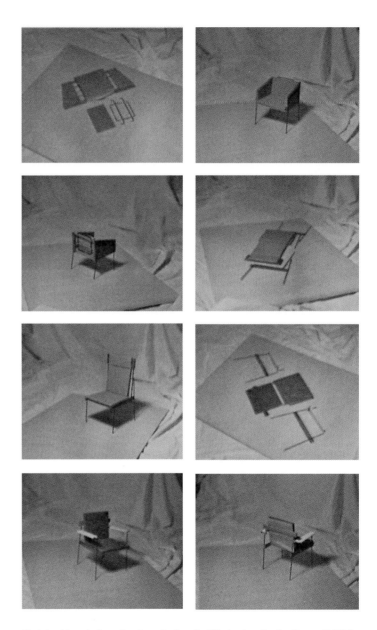

Models of knock-down furniture designed while freelancing for Sears. (1957)

STARTING MY DESIGN CAREER

(IIT), where the design program had become accredited. The night program at IIT didn't offer industrial design courses, so I pursued a degree in art education. It took me about four or five years to get a master's degree. My thesis and graduate project were entitled "An Exploration in Fragmented Wood Forms."

I also found freelance work, designing toys and whatever other design work I could find. Then a friend told me about a design firm that was looking for people. I showed up the next day, and all of a sudden, when I walked in the door, they no longer needed any help! It seemed I was the answer to everyone's employment shortage problems! This was 1956, and, of course, it was a racial thing.

I pounded the streets until September and never found a job. I had a few classmates working for Sears, Roebuck & Company who told me that there was an opening for an industrial designer. So, I tried Sears and interviewed with Carl Bjorncrantz, the manager of design. My furniture background caught his interest. Carl was very kind and felt for people who struggled. He would go the extra mile to move away obstacles for people—he had a couple of people on his staff who survived polio. One guy had a complete leg missing. Carl was just a very kind human being. A person's design talent had to be there, but he gave you a chance to show what you could do.

I completed the application and took a battery of tests from their personnel department. On the next visit, Carl looked kind of stressed and said, "Chuck, I have to tell you. It turns out I can't hire you. I know you need an explanation and I'm going to tell you like it is: Sears has an unwritten policy against hiring black people." Of course, this didn't put Sears in big trouble, because in those days there were no laws against discrimination. But Carl did have the guts to tell me, which was rare. I admired him for that. Even though he couldn't offer me permanent employment, he offered to hire me on a freelance basis.

So I started working freelance for Sears from our apartment. Because of the strength of my furniture design and despite the fact that Carl had about five designers who concentrated on furniture, he wanted me to provide them with additional design ideas.

Around that time, I got a call from Henry Glass. Since he knew me as a student, he knew the quality of the work he would get. He didn't even have to interview me or look at my portfolio. He told me that he and his partner, Lou Huebner, had decided to go their separate ways—Lou to pursue architecture, and Henry, industrial design.

While Henry was an architect, he had a great reputation for furniture design, so he decided to focus on furniture, before expanding into other products.

Henry said, "Would you consider coming over and going to work?" I said, "What time do you close? I'll be there."

Henry Glass was one of my most influential mentors in addition to being one of my professors. He was also my first industrial design employer and a great friend. (2001)

THIS PAGE AND NEXT: Fragmented wood form sculptures from my master's thesis project at the Illinois Institute of Technology's Institute of Design. (1963)

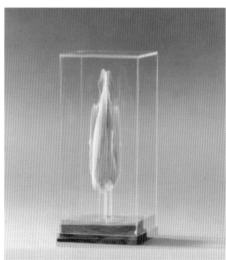
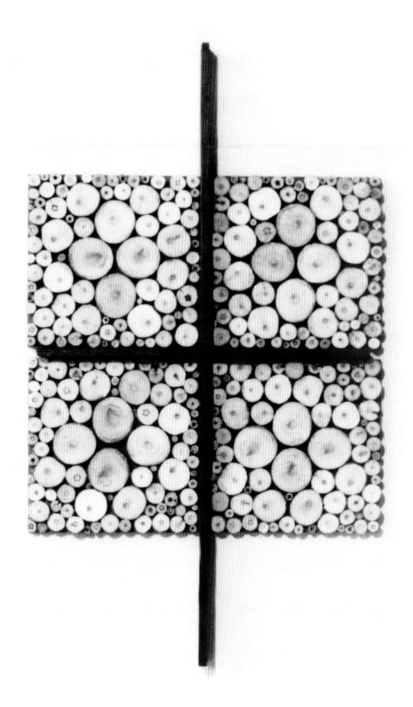

STARTING MY DESIGN CAREER

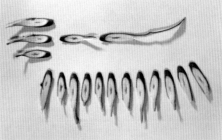
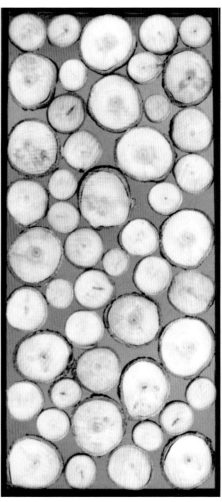

He laughed and said, "Tomorrow's soon enough."

So, I worked for Henry by day and Sears at night. Every two weeks I'd find a way to get out to Sears to deliver the work to Carl at the headquarters on Chicago's west side at Homan and Arthington. I did that for three or four months until one day Carl notified me that his budget for freelance designers was running out and he could no longer keep me on as a freelance designer.

Henry was a real godsend—the real key to getting me going as a designer. I now had someone looking over my shoulder, checking my mistakes and teaching me. I learned that it's not only about expressing a concept, but how to successfully get it to the marketplace.

I worked primarily on furniture design, but I learned a great deal from Henry about detailing and drawing. As a result of working with him, my drawing skills became professional. I drew up most of his ideas from his thumbnail sketches, and I would add my own thinking. When he went to a meeting, he'd take along these drawings to the client, and once they made a selection of the concept, he'd bring it back and I'd detail it. I drew the furniture, full size, on a long drawing board and Henry would look over my drawings. Because the drawings were full size, he could see problems and identify them by putting his finger right on where it was.

From Henry, I also learned a lot about the business, watching how he dealt with clients and how he prepared for meetings. I learned about the materials and production of furniture, managing client relations and meeting deadlines. I was designing for the real world—products that were to be mass-produced.

Designing something to be made in the hundreds or maybe thousands of units required more advanced thinking. I learned that designers must understand the appropriate selection of materials and available production processes. A lack of materials and production knowledge can result in slower than necessary production rates, and choosing less appropriate materials or processes may make the product too costly to bring to market. The designer must also make sure that costs of tooling and the amount of materials are kept at a minimum. If the process is too complicated, production is likely to be slow, and if it's too delicate, it will be prone to damage. All of this directly affect the cost of the finished product to the consumer.

I enjoyed working with Henry, but I began to get itchy. I was young and eager to know more, to be capable of designing anything. Also, Janet and I were getting married that August.

A little side note here: Henry's office was in the tower of the American Furniture Mart at 666 Lake Shore Drive, which was then the tallest building in the area. At the time, Janet worked for the Institute of Psychoanalysis on the top floor of a building on Michigan Avenue. There were no buildings blocking the view between us, so we'd

On a date with my wife-to-be, Janet, at Tito's Hacienda on South Halsted Street in Chicago. We were both students in the city. Janet was living at the International House on the University of Chicago campus. (1958)

STARTING MY DESIGN CAREER

This clothing storage unit for Saginaw Furniture Company was a pull-out system to store hanging items and shoes by combining a chest of drawers with a closet. I did most of the detailing on this piece after the concept was initiated by Henry. (1957)

put a piece of paper in our respective windows to signal we were ready to go to lunch.

After working with Henry for some time, I started getting calls from people all over the city who were aware of my design abilities. I guess they figured if I had stayed at Henry's for three months and didn't bite anybody, they could hire me. From that point on, I didn't have to look for any jobs; people called me.

Marshall Field's ad for TV trays designed for Quaker. (1957)

STARTING MY DESIGN CAREER

I answered one call from an aggressive young designer named Ed Klein. He had been on the design staff at Motorola but had a one-man office, concentrating in electronics. Ed offered an opportunity to fulfill my desire for experience designing products other than furniture. I still remember how difficult it was for me to find employment and was very excited at this opportunity, so I gladly accepted his offer.

I grew a lot at Ed's place. I began to learn about manufacturing in metals and plastics in greater depth. We designed a lot of radios, televisions, pianos, electric organs, carpet sweepers and lawnmowers, among other products. I did some gas-powered lawn mowers for an Amish manufacturing company and ironing boards for a company in Indiana. We also designed barber chairs and trivets for hot dishes, and private brand products for Montgomery Ward.

In the mid-1950s the television industry really began to expand and most people were buying floor consoles. I designed the cabinets, which were really pieces of furniture—beautiful pieces of cabinetry. Even though portable TVs were not in strong demand, Hotpoint sold portable units with metal cabinets, and we were retained to design these units.

Around 1957, Zenith began producing the first Trans-Oceanic radio, which was about the size of two shoe boxes. When that radio hit the market, RCA decided to start producing its own version of the radio. Since Ed was pretty aggressive, he sold them on our design services and we got the job designing RCA's version of the radio. RCA's strength in the marketplace and the quality of our design helped their radio take significant market share from the Zenith model.

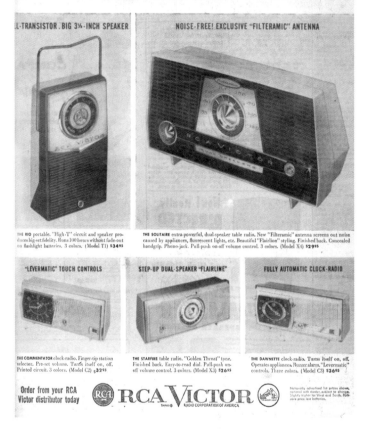

Klein was retained to design consumer electronics for RCA including these early transistor radios. (1958)

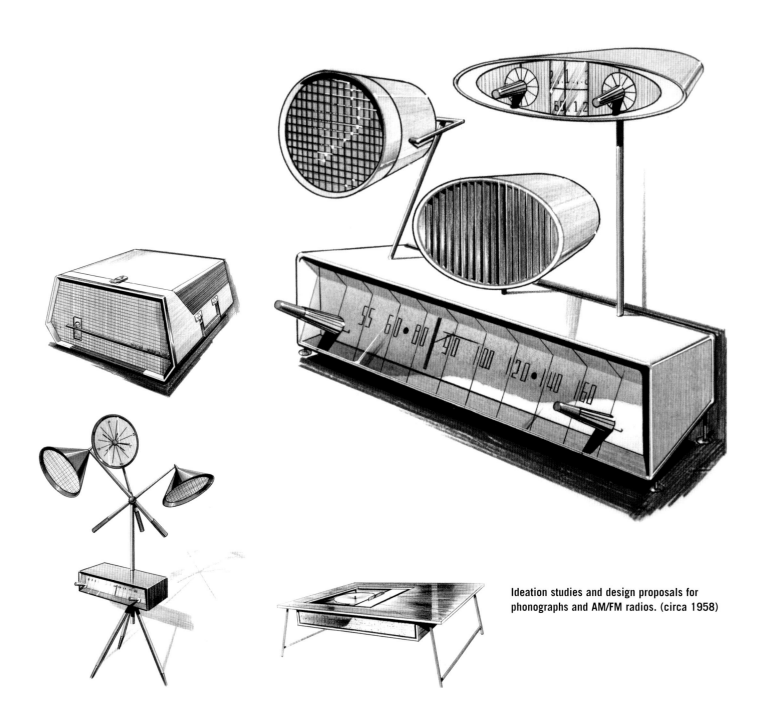

Ideation studies and design proposals for phonographs and AM/FM radios. (circa 1958)

STARTING MY DESIGN CAREER

AM/FM table radios for Allied Radio Co. and Western Auto Supply. The units used plastic enclosures with plastic grilles. Below is a display of all the parts used in manufacturing one of the radios. (1957)

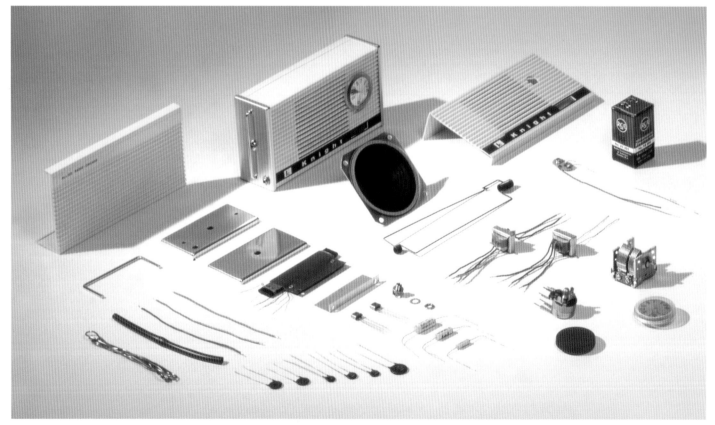

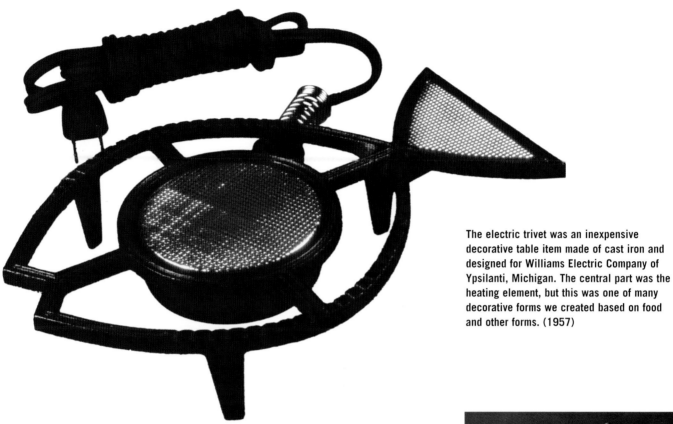

The electric trivet was an inexpensive decorative table item made of cast iron and designed for Williams Electric Company of Ypsilanti, Michigan. The central part was the heating element, but this was one of many decorative forms we created based on food and other forms. (1957)

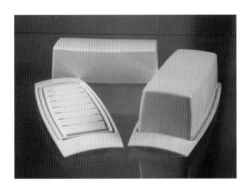

Butter dishes, my first mass-produced product, a low-end housewares product made of injection molded polyethylene, designed to be attractive enough to be pleasant on a table setting. Designed for Victory Manufacturing Company. (1957)

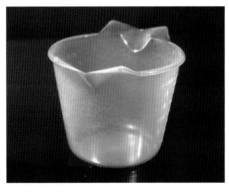

This measuring cup, made of polyethylene for flexibility and damage resistance, was one of my earliest plastic designs. The cup was transparent to allow the user to see its contents. (1957)

Flexible plastic pails with lids made of injection molded polyethylene. The lids had an integrated hinge that developed during the molding process. (1957)

STARTING MY DESIGN CAREER

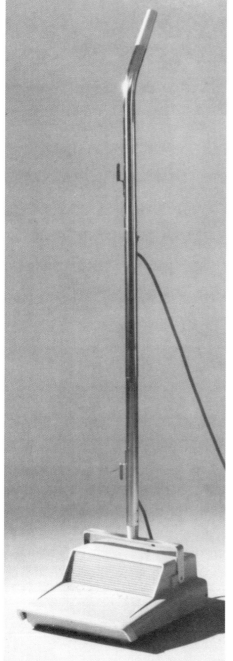

OPPOSITE TOP: Electric food warmer made of cast aluminum with wooden handles by Williams Electric Company in Michigan. (1957)

OPPOSITE BOTTOM: Upright piano, designed for Grinell Brothers Piano in Detroit, Michigan, made a definitive statement in any home. The product demonstrated my strength in furniture. (1957)

OPPOSITE RIGHT: This electric carpet sweeper was probably the first electric-powered carpet sweeper for Davis Sweeper Company. It was one of my earliest products. Its plastic housing was also an early innovation as most of the products of that era used sheet metal. (1957)

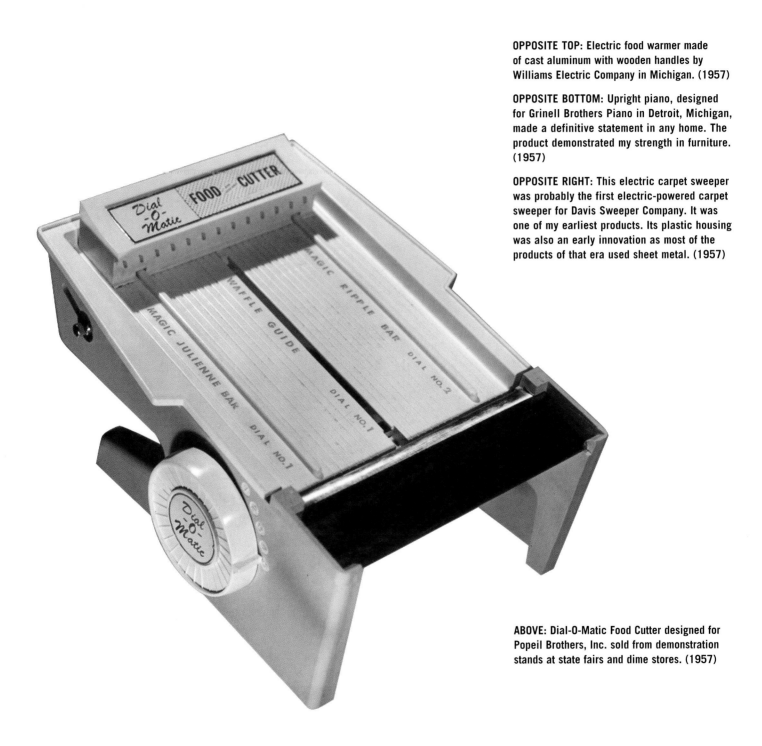

ABOVE: Dial-O-Matic Food Cutter designed for Popeil Brothers, Inc. sold from demonstration stands at state fairs and dime stores. (1957)

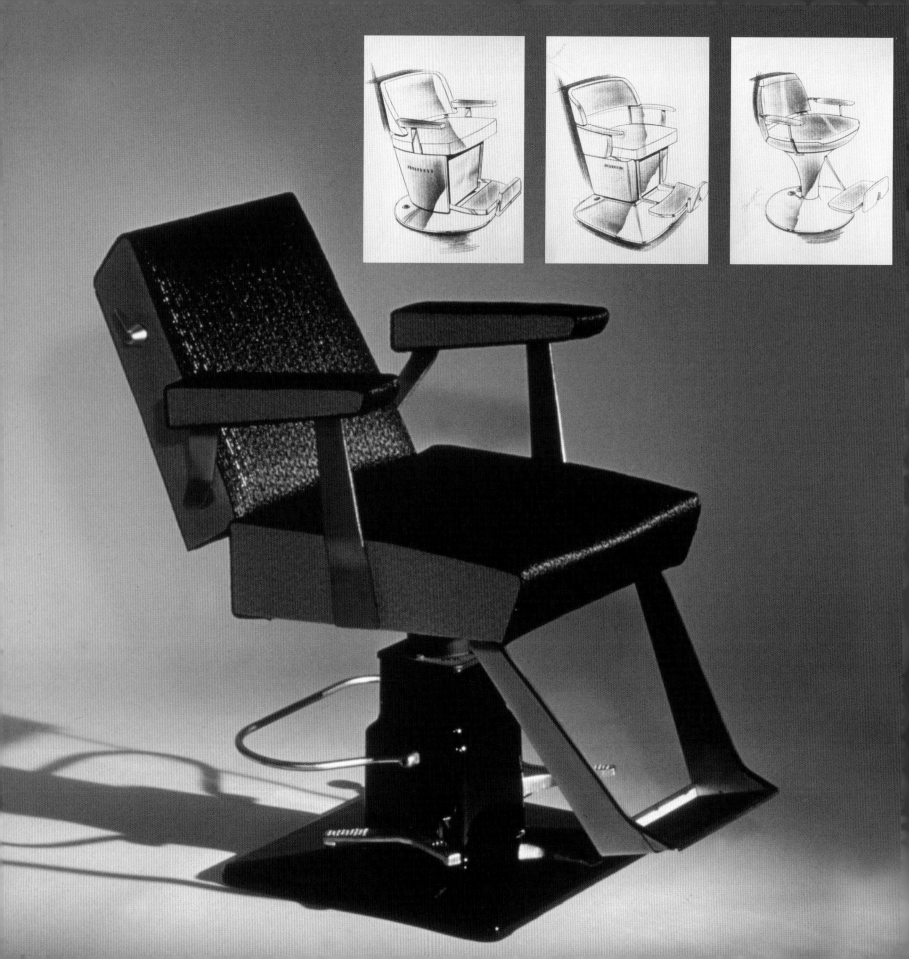

OPPOSITE: A beauty salon chair model and barber chair concepts designed for Belvidere Beauty Products in Belvidere, Ilinois. (1957)

RIGHT: This truck grille was designed for Diamond T Truck Co., which enjoyed a quality reputation and a strong, masculine image. The grille was the primary style element on the face of their trucks. This was one of a group of concepts that we submitted. (1957)

Ed Klein & Associates was just one of many design firms clustered along Michigan Avenue around the Wrigley Building, banging out the great design that helped define the look of domestic life in America during that period.

Many guys from the back rooms would see each other for lunch. We couldn't afford the Billy Goat Tavern, which was frequented by writers from the nearby *Chicago Tribune* and *Sun-Times*, and the political who's-who of Chicago. We'd brown-bag it with sandwiches and sit along the banks of the Chicago River. We talked about where we could get five more bucks a week for our work. If you wanted a pay raise, that was how you got it—by changing jobs. These sessions by the river provided our unofficial classifieds for job openings. By the way, there were only a few women designers at that time. Like many other professions, industrial design was a good ol' boy, Caucasian male domain.

Although I was second in command at Klein's, I felt strongly that it was time for me to move to a bigger-name firm. I had been on waiting lists at Raymond Loewy Associates, Dave Chapman Associates and Mel Boldt & Associates, which were all big design firms in Chicago with international reputations. But, I didn't get called by any of those companies. I did receive a call from another designer telling me that Robert Podall Associates was looking for a designer with my skills. I contacted Podall and soon moved over to his design firm in the Lincoln Tower Building, right around the corner from Ed Klein.

STARTING MY DESIGN CAREER

Podall's was beautiful place to work. Bob Podall was a pleasant guy to work for, and he had great drawing skills. With Ed Klein, I saw how he operated, but didn't learn a lot about the business. With Podall, I did a lot of the business work in addition to drawing. I estimated jobs, presented design projects to clients, and interviewed potential employees when they called. I estimated jobs—how long it would take for us to get the work done—which puts some stress on you. If you promise the boss you're going to get it out by a certain date, and he draws up the contract and it doesn't come through, the company and you can be in big trouble.

THIS PAGE & NEXT: Renderings and models of early transistor radios for Western Auto Supply, Allied Radio and Montgomery Ward. This was an era when transistor radios were coming into a strong position in the market. These were styled to bridge from vacuum tubes to much smaller transistors. We thought the customer needed a slow introduction into transistor radios. Because of that we didn't try any ground-breaking visual images. (1957)

As part of our consumer electronics work, we designed some of the first transistor radios made in Japan and exported to the U.S. The Hinners Gallenic Radio Corporation, based in New York, contracted Podall's to make designs and full-scale models that in many cases looked better than the production units. The manager for our client would select the design and we would send control drawings and a model to Japan for manufacturing.

On one of the models that we shipped, the pressure tape holding one side of the nameplate popped out of place. Since it was very important for the Asian companies to do exactly what was asked for, the company actually produced and shipped several hundred of the radios with the nameplate sticking out! This was a costly mistake, but since the control drawings were accurate, the manufacturer had to eat the cost and send a team to the U.S. to correct the mistakes.

Bob always consulted with me on what I thought. My role was to see that we worked fast enough to get the designs out on time for us make some money on them. I had to make sure that the quality of the work was excellent, while also designing, so I was sort of a player-coach. I stayed at Podall's for a couple of years.

We did a lot of the same type of work at Podall's as we did at Klein's, including a slot machine for Bally's and a small line of tabletop appliances for Casco. The line included toasters, electric skillets, toaster ovens, steam irons and other products, which would be produced by major manufacturers and sold under the Casco brand.

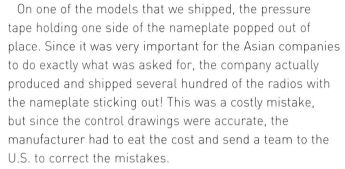

STARTING MY DESIGN CAREER

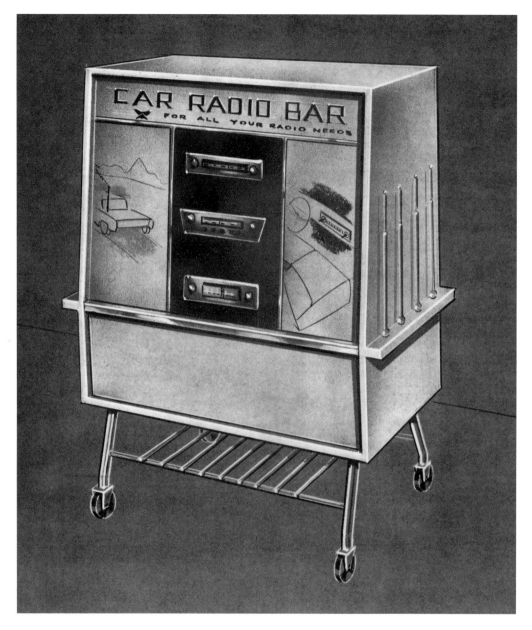

These renderings were for a point-of-purchase display. Automobile radios were frequently an aftermarket addition or upgrade at the time. The systems would have been functional, allowing customers of auto parts retailers to test the radios prior to purchase. (Podall)

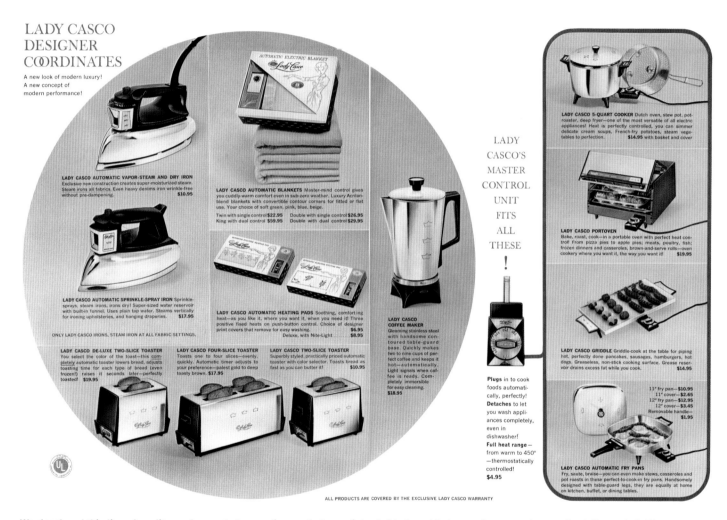

We developed this line of quality grade countertop appliances that were intended to be sold at one price everywhere and not discounted. The Lady Casco brand didn't succeed in the marketplace primarily because it was undersold by less expensive products. (1957)

STARTING MY DESIGN CAREER

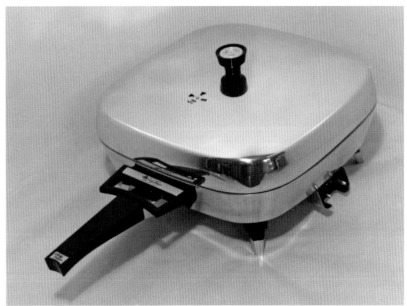

THIS PAGE AND NEXT: Products from the Lady Casco products line included a steam iron, an electric fry pan, and the all-purpose Chef-Mate (opposite). (1957)

STARTING MY DESIGN CAREER

OPPOSITE: Proposals for coffee beverage dispensers. Designed for Hava Java, they were proposed as built-in additions to homes and offices. These preceded the strong growth of the vending machine market. (1958)

RIGHT: Slide projector made of die cast aluminum for Sawyer's Inc. (1958)

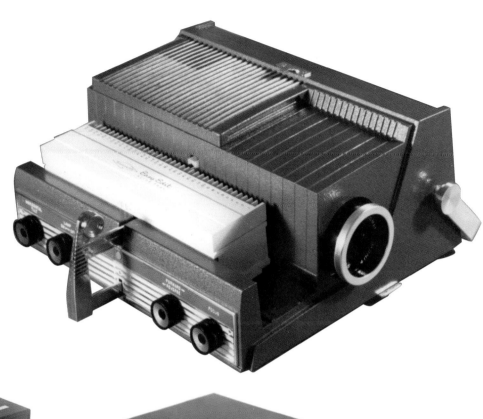

Closed and open views of injection-molded modular stackable slide tray storage units designed for Sawyer's Inc. (1958)

Actually, the most visible design I'm probably known for came while I was working for Bob, and it was almost an accident. Podall's was retained to design products for Sawyer's Inc., which manufactured slide projectors, slide trays and other photographic equipment.

One day a project came in from Sawyer's—It was a 3-D viewer—and Bob just put it on my desk and said, "Here's another product." There was nothing special about it. It had been around for quite a while, so I didn't invent it myself. Up until that point it was made of a hard, dark plastic called Bakelite, which involved a very slow compression-molding process. As a result, the product was costly because it took so long to make.

Sawyer's Inc. retained Robert Podall Associates to design their products during the late 1950s and early 1960s. I was assigned the project to update the View-Master. The design parameters required a lower-cost item that incorporated many new features and engineering improvements. One of these new features was a disc with pre-selected slide-type visual stories. Additional stories could be purchased separately.

My contribution was to design it for a different process, injection molding, that could produce units 10 times faster and would reduce cost considerably. The projected volume was high enough to support the tooling costs, while keeping the amortization costs low for each piece part. This made it possible to get the finished product to the consumer at the target selling price.

I changed the visual esthetics of the viewer, giving it an appearance that was appropriate for its time. I changed its form factor and color to match the rest of the Sawyer's product line. The project really wasn't very big. It was something that came in and was out in two weeks! We made models of it and it was put into production. The View-Master was one of my most recognized designs and longest-selling products I designed.

The View-Master was later bought by GAF and marketed as a toy. And it sold and sold. I think its success and popularity were due to the excitement for kids to see photographs in three dimensions. There was not as much television around then, kids didn't have TVs in their rooms as they do today, and they had no personal computers. If their parents were watching something the kids didn't like, the kids were out of luck. So they could take their View-Master and see a story, or use it to keep entertained when traveling in a car. It just happened to hit the American lifestyle at a time when it could fill an entertainment need.

Over time, I don't know what happened to Sawyer's. The tooling for the View-Master was sold to GAF, and the colors were changed to give it more flair and appeal to children. I designed it in the late 1950s and eventually it was sold to Fisher-Price®, but only recently, after almost 40 years, was the form factor changed. The View-Master still survives despite all the competition from other high-tech products and toys because of the appeal and fun of its little bit of three-dimensional magic.

OPPOSITE: The View-Master's shape was developed to provide ease of use and comfort. The original color was beige, in order to make the View-Master subtle and compatible with other photographic products in Sawyer's product line. Later, the color changed when the View-Master ownership was sold to GAF. (1958)

STARTING MY DESIGN CAREER

In a small firm like Podall's and Klein's, it was not uncommon for us to work well into the night. Ed Klein was notorious for trying to get all he could out of us. At Bob's, we'd work long nights and occasional weekends. Bob was pretty demanding too, but he also expressed his appreciation for our hard work. One night, I was going out to dinner with Jim Aurand, a colleague and one of the most creative people I've ever met. We were meeting our wives to celebrate our wedding anniversaries. At the time, our office was also under a lot of pressure to meet a deadline, and even though it was customary for everybody to work around the clock when we were on deadline, Bob gave us the night off and gave us the money to celebrate.

Still, it was pretty rough in the trenches of a small shop; where if the boss didn't have enough money on Friday to make the payroll, he might ask a guy not to come back on Monday. I wanted a place where I could pull the cord out of the wall at five o'clock, close up and go home. At that time, some of my friends had jobs in the corporate world, and I began to want a job like they had, where I could get vacation pay, hospitalization, even unemployment insurance.

Game for the ReRon Sales Company. These animal characters are pieces for a game that I developed as a freelance worker in collaboration with Jim Aurand, the most creative designer that I ever met. (1966)

Janet and I, along with friends John Pryer and Don Patton, built this A-frame summer cottage on a stream near Vandalia, Michigan. John and Don had been schoolmates of mine at the School of the Art Institute of Chicago. I designed the house, structurally and esthetically, as a place to get away from the city. We searched a radius of 100 miles around Chicago trying to find some land that we liked and could afford. We would travel on weekends, scouting land and asking people for places that we could purchase. We happened upon someone in Michigan who owned this property and quickly sold it to us. The house took three years to complete and was entirely built by the four of us. The glass for the front of it was cut in Chicago and we transported it to the site. Through the blessings of God, they fit. We installed a maple hardwood floor and all of the appliances and furnishings. Built over weekends, the house faces a beautiful stream and sits on two acres of land, one cleared and one wooded. The back of the house faced Calvin Center Road. We sold the house to an artist in the late 1960's as we began to have children. For me, the great pleasure of the house was in building it. (1963)

"I first met Chuck when I was a buyer at Sears and he was a manager in the laboratory. I asked for a design on a guitar case I was developing, and he came through for me on that one. Later, we traveled together in Europe, when I was heading up our European buying operations in Frankfurt."

~Carl Boelke, now retired and living in Seattle

WORKING AROUND THE WORLD

In 1961, Sears finally did come through. I got a call from Carl Bjorncrantz, saying, "Well, Chuck, we can hire you now." I said, "Thanks, but I have a good job," which was true. I was doing a lot of interesting work, a lot of home appliances, electronics and other products, all over the board. Carl said, "At least come out and talk to us." In the interview room, I remembered all the nights I had to work at Podall's and how the business was small and fluctuated, and I thought that I could use a less volatile ride. Meanwhile, Janet had become pregnant a couple of times, but we weren't successful in having children for quite a long time. We lost three. So it wasn't easy for either of us, especially Janet.

Carl offered me a job. As it turned out, I was the first black executive Sears ever hired in headquarters. They had porters and a couple of guys in the cafeteria and maybe a stock room guy—but there were no black secretaries and no other black person on what they called the checklist level of employment, which was the executive level. During my interviews, I had to take a battery of psychological and interest tests that Sears used to fit people into various levels of their corporation. I also was grilled by four members of personnel.

In the end, I got the job, and was happy to be at Sears. I enjoyed it, although I encountered a lot of racism, which came from every direction. Fortunately, by this time I

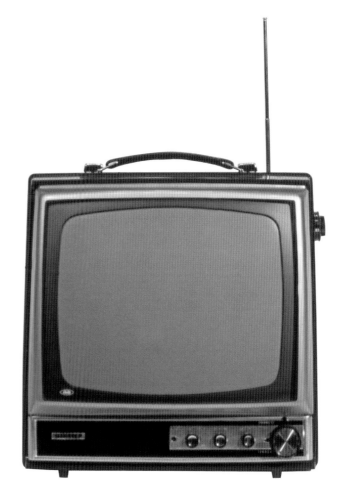

This portable television was the first Japanese-made black and white TV to enter the U.S. marketplace. It was simple and clean in appearance, with only basic features, and was marketed to the low end of the price range. The TV had very good quality and dependability, which helped reinforce the image of the Silvertone (Sears' private label) brand. (1963)

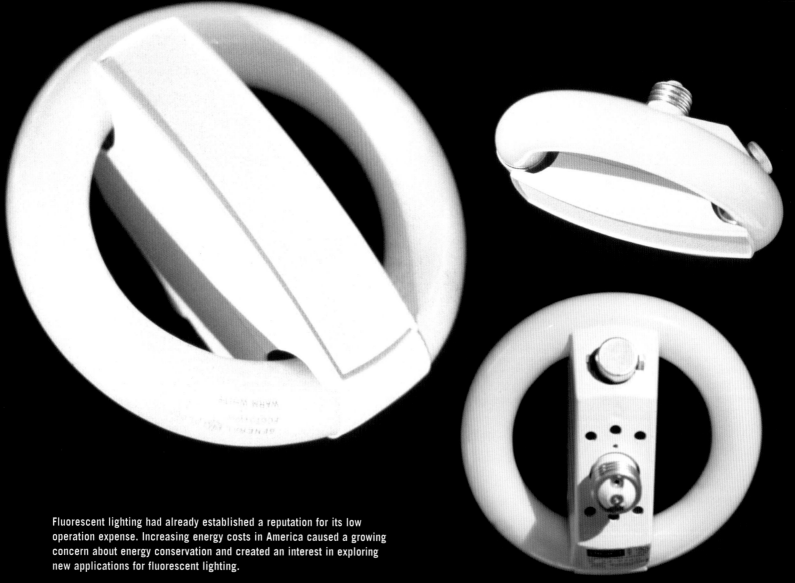

Fluorescent lighting had already established a reputation for its low operation expense. Increasing energy costs in America caused a growing concern about energy conservation and created an interest in exploring new applications for fluorescent lighting.

I received the assignment to design a low-cost adapter that would allow a customer to use a circular fluorescent lamp in a conventional screw-in socket. The adapter would make it possible to replace regular incandescent light bulbs in table and floor lamps, as well as in ceiling fixtures. The science was unsophisticated, with little research and development having taken place up to that time. Nevertheless, the need for an adapter existed. We reviewed state-of-the-art and off-the-shelf components and made choices based on performance requirements. The final components were creatively arranged to fit inside a small enclosure. The enclosure, or housing, had to be lightweight and small in order to obstruct as little light as possible. Also, the unit had to look attractive, hold the fluorescent lamp securely, and provide easy access for replacement of internal components. The solution was an injection molded plastic product that was granted a U.S. design patent. This design set the standard for similar products to follow. (1970)

knew how to handle it, from college in San Francisco to the military and in the workplace. At Sears, most of the racism came from people in the company like buyers and sales managers. My fellow design workers, by and large, were accepting. In fact, I think they were relieved after they found out I wasn't going to destroy their positions. I had by this time developed a strong reputation around Chicago as a capable designer.

The transition of moving from a small firm to a big corporation proved to be smooth in terms of performing industrial design. In fact, I was probably head and shoulders above most of the designers at Sears in terms of my ability to do the work. I had a rougher time figuring out how to weave my way through the corporate stuff, the bureaucracy, and learning how to play the game—which I don't think I ever really learned. People in corporate America are really not any brighter than people in small companies—in fact, in many cases, they don't think at all. They just do their little part of what they're supposed to do, and they don't go beyond it. So I wasn't in awe of the competition there, at any level. If anything, I had some disdain for some of it, because they flaunted their power and their positions, even though there was so much mediocrity.

Initially, I went to work as a designer and over time was promoted to group manager in the design section. My design activity was centered on consumer products. I designed all major kitchen appliances, lamps, wall systems, casual and upholstered furniture, stereo systems, executive chairs, electric manicure sets, leaf and grass blowers, fish tanks, luggage, toys, toilets, camping gear like tents, and exercise equipment.

I also designed a screw-in adapter for a circular fluorescent tube, which, today, is very plentiful. It's

Console televisions were the big-screen televisions of their time, representing a possession of pride and a piece of furniture in many homes. Since most appliances were free-standing, the televisions were designed to work nicely in households of the time, keeping pace with trends and interior decorating demands of discerning consumers. (1962)

WORKING AROUND THE WORLD

Electronics was part of my design experience when I became a full-time employee at Sears, Roebuck and Company. Shortly after starting, I was assigned to a project for H.C. McCoy, the senior buyer of television sets at Sears. The company was increasing its foreign suppliers, and Toshiba was selected as a primary electronics manufacturer for Sears. Together, we designed the first Japanese-made, black and white TV to enter the U.S. marketplace.

Dinner in Tokyo with representatives from a radio manufacturer with Sears personnel Dick Shultz, Kent Venema, Chuck Shimatani. (1978)

intended for low-cost lighting in places like hallways. Sears won a patent for the design of the product, but as usual, I didn't get any royalties because it was normal practice for employee to sign an agreement giving Sears ownership of work created while in their employ.

As I mentioned, I was fortunate to travel and work all over the world. We often traveled overseas in search of manufacturers to produce products that weren't available through American manufacturers. On one occasion in Europe, we were trying to find a company that had the equipment to make furniture of a particular style. Eventually, we went to a town in northern Italy that was near Venice and found a manufacturer. We toured the plant and became acquainted with the people there and discussed some of the issues like shipping and production rates, then went to lunch with their young sales manager. He told us he had lived in the United States—he had relatives in the moving business—but that he didn't like it there. He said the thing he disliked the most was that everybody was treated the same, so he returned to Italy! I was taken aback. I mean, that really gave me a jolt.

On another trip, I had been working with some suppliers in Tokyo who produced personal care items like hair curlers, and they took us fishing one weekend in a town about five hours south of Tokyo. I didn't know anything about fishing in Japan. They took us out in a small, very tired-looking boat with a diesel engine that went "pop-pop-pop." I discovered that in their style of fishing, the Japanese used a line without a fishing pole. You held the line with your fingers; some feet down, and the line ran through a little canister that looked like it should hold 35mm film. Instead, the canister had perforations and the guide had filled each one with ground-up fish parts. About four or five feet below the canister were five or six small hooks. The idea was to jiggle the line with your fingers to move the canister and create a cloud of flavor in the water that attracts the fish. In my mind, I could see a big fish coming along and just yanking my finger off! It turned out that the fish weren't that big, but I wasn't really anxious to catch one anyway. I didn't jiggle my line too often! That was an interesting experience for me.

WORKING AROUND THE WORLD

Early on, some people expressed concerns that Sears might be afraid to send me on a business trip because of how I might be received (as an African American), even though they knew darn well that the company I was visiting wasn't about to turn down a million dollars' worth of business just because of me. So there was a lot of hocus-pocus stuff. Over time, I became more accepted, although until the day I left Sears, I was always reminded that I could not take my guard down, that I was in a hostile environment every day. Every day! That mind set was so ingrained at Sears, in its employees, and probably still is.

Occasionally, Sears asked employees to field-test some of the merchandise. On one of those occasions, I volunteered to test tires. After dropping off my car to have the tires installed, I had to go down an alley between two buildings to get to my office from the garage. As I was walking back to the main entrance, I looked up and saw the security police with lights flashing.

They stopped me and said, "Hey, where are you going?" I said, "I'm going to work." "Where do you work?" they asked. I said, "In this building." They said, "Oh. You do?" So, I showed them my ID, and they said, "Fine. Some guy looking out his window told us he saw a black guy walking between the buildings, and we had to investigate." This incident happened more than 10 years after I started working at Sears. Apparently being neatly dressed, wearing a suit, didn't matter as much as the fact that I was black.

Another incident happened some time later when I got a call from Henry Glass, believe it or not. Now, I had been at Sears maybe 10 years and hadn't worked for Henry for more than 15 years. He said, "Chuck, are you leaving Sears?" I said, "This is the first I heard about it. Tell me where I'm going."

He said, "Oh, I just got a call from this guy at a detective agency who wanted to check your employment history." I

A weekend in Portofino, Italy during a trip with my buyers.

Dinner In Osaka with Sears vendors.

said, "I don't know a thing about it." Henry said, "Well, if you're going to leave, let me know, because we'd be glad to have you."

So I walked right into my manager's office and asked why this detective agency was checking on me. He said, "How did that happen?" I replied, "I don't know, but somebody at Sears has obviously directed him to investigate me."

My manager said, "That has to be some kind of mistake." I said, "I don't like it," and walked out and started asking questions. I couldn't find anybody else who had been checked except a couple of women in the home economics lab. It was just them and me, the only black guy. It turns out management was investigating us because of the theft of some materials from the lab on the floor above where I worked. Well, about 20 minutes later, I got a call from my manager, who said, "Chuck, I really want to apologize. I don't know how it happened. But we want to show you we really appreciate your being here, and we're going to give you a pay increase." Well, it was the largest increase I ever had!

I found out later it was two white guys, technicians, who were really doing the stealing.

There were always incidents like that. Every day somebody said something out of order. In most cases, they didn't even know what they were saying.

On tour with my hosts in Japan.

Several years later, I recall sitting in the office of my immediate boss, with whom I got along with fairly well. He depended on me quite a bit for information because he didn't have the experience I had. When he got into a tight place or had to make some decision regarding design, he'd ask me about it. We were talking casually this time, and he said, "Oh man, there's a nigger in the woodshed somewhere here." I didn't take this in a hostile way. He probably said it every day, just not in the presence of African Americans. Overall, the racism at Sears didn't let up as the years went on. Never did. It was so ingrained. And I don't think it was necessarily just Sears. It was America. I'd ride to work from the South Side with Bob Johnson, an African-American colleague, and we'd stop outside the Sears building and pretend we were putting on gas masks, preparing to go into a hostile environment.

I also remember when Martin Luther King, Jr. was killed in 1968 and rioters started setting fires around Sears headquarters, all these Caucasian employees were caught inside the complex on the ghettoized West Side of Chicago. People were getting out of there in any way they could. One guy even dressed up in a nun's habit! Really. So he could get out of there and get back to his white suburb. He obviously was terrified.

Eventually at Sears, upper management began dividing the company into all kinds of small groups. In most cases, they made it less functional, less productive. They just added more management to what was already there.

In 1979, I was taken out of the design department, leaving two design managers, and promoted to Group Laboratory Manager. The purpose was to provide technical support to a specific merchandise group. We worked from the Chicago headquarters, but we serviced Sears all over the world. I was the manager for a merchandising section, which included automotive, sporting goods, tires, batteries, candy and other foods. I didn't consider it my forté, but it was considered a promotion and I stayed in that assignment for two years, despite the fact that I found the job very unpleasant. It was out of the realm of design where I could perform a service and see a result. As a manager, I was in a political arena where I didn't have a hands-on role in developing products.

In the early 1980s—we were now in the Sears Tower downtown in the Loop—Sears started downsizing. It brought in an executive from New York, with the specific purpose of downsizing the lab. He began doing things to try to get people to quit. It was pitiful. He handled it in a very crude way, not supporting his people and hanging many of them out to dry. Sears whacked away, and eventually got rid of thousands of people over 10 years.

Seeing this, I wanted a change in my job assignment. In 1983, one of the co-managers of the design group became ill and he couldn't perform his duties. I took a chance, and told the head of the Lab that I really didn't like my current assignment. I asked to return to the design group, to assume the ill man's responsibilities—and he actually let me do it. I moved back to the design group and worked for two or three years as co-manager with the other fellow. When he decided to retire, I assumed responsibility for the entire design group. And I stayed there until the whole section was closed in 1993.

Unfortunately, Sears management had started to care less about design and quality. This began just about the same time they undertook downsizing. Management decided that it would drop quality control down to very few people. It started depending on the product suppliers to provide design, as it still does today. Talk about putting the fox in the hen house! Many buyers had no idea what

OPPOSITE: Janet and our son Charley on a visit to Hawaii. We met there as I was returning from business in Japan.

WORKING AROUND THE WORLD

Standing in pecking order was protocol in Japan. In this photo at the Maruzen Manufacturing Company, the man in the middle was the highest ranking member of the Sears contingent, and the rest of us were in order of power. Sears people were on the left and Maruzen employees on the right. National Manager, Buyer, Designer, Manager of Sears' Tokyo office, and the Buyer for the Tokyo office. (1970)

they're looking at. So, they were completely at the mercy of their suppliers, who gave them the royal treatment. The buyers, romanced by their suppliers, often overlooked a lot, which impacted quality control.

The emphasis on profit margins and markup got so strong that the attention to detail and quality was lost.

Sears ended up selling much of the same merchandise that other retailers sold—the same product but with a different label on it. Sears had been known for its quality and value, but these products were no better than anyone else's. There were few places you could go and buy merchandise of high quality for the same amount of money, but today that's no longer true. You can go to a lot of places and buy the same product for less.

The company walked away from many areas where it had a complete corner on the market. Sears was known for building materials and Craftsman tools. It was dominant in tires, batteries, furniture and paint. It even had a strong position in the men's clothing industry, believe it or not.

When I retired in 1993, I felt it was a sad day because of the direction of corporate America, and Sears in particular. The company had decided it wouldn't need any more in-house design people. Before I left, I saw about 22 design people go. It was a tough experience, and I was the last designer to leave, which was kind of ironic, since they had been so slow to hire me in the first place.

LOOKING BACK

I wish I knew how many products I have designed in my lifetime: probably at least two or three thousand. Now, when I say design, I don't mean there are that many out there that exist as living products, but I have designed that many things in concept. A lot of designers like myself, who were in the back room working on the drawing boards, were paid to generate ideas. We were expected to produce at least one or two ideas every hour and sometimes even more than that if you were doing quick sketches. My guess is that close to 750 of my designs were manufactured.

Looking back at the major turning points in my career, I'd say the first was when Henry Glass gave me a job. Even as a teacher, he gave me the confidence that I was really as good as everyone else, or could be, despite being black.

When I went to work with Bob Podall, I felt in my gut that I was truly a designer—that I knew enough about the field and was ready for the world. I managed his office and he trusted me with a lot of responsibilities that no one else had. And I felt confident in working with almost any manufacturing process—stamping, metal processing, die-casting and all the plastic processes. I also have to give Ed Klein a lot of credit, because it was at his firm that I started to learn to design products other than furniture.

After moving to Sears, it all crystallized. I began to really understand all the economics that interact with design, along with the politics. During those 30-plus years, I became a very strong designer, without exaggeration. I felt I could hold my own with anybody in the world. There were probably very few people who had as much real experience and knowledge that I had.

WORKING AROUND THE WORLD

I designed hearing aids to fit into the temples of eyeglasses and sunglasses. My goal was to make the make the hearing aids unobtrusive. Initially, hearing aids were large enough to fit in a pocket, then aids were design to hang behind the ear and we work on reducing their size to these which were part of the eyeglass temples (below). The temples were close to the shape of those on regular glasses. (1974)

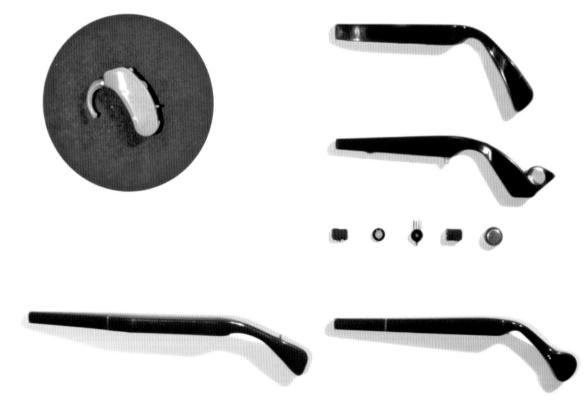

Single-record child's phonograph with a "Winnie-the-Pooh" theme. The record player also came in other versions with Disney characters that danced when the turntable moved. (1979)

WORKING AROUND THE WORLD

RIGHT: The desks and chairs were designed for home and small offices and manufactured in Jasper, Indiana which was a hub for wood products in the midwest. (1975)

BELOW: Modular shelving system for Sears, made by Van Pelt Manufacturing Company of Antwerp, Belgium. Offered in teak, walnut, veneer or white finish. (1975)

LEFT: An early U.S. introduction of flat panel furniture which later became very popular. (1976)

BELOW: This wall unit was made exclusively for Sears in Belgium by Van Pelt Manufacturing Company. Walnut tree logs were exported to Europe to supply material for the veneer. (1975)

WORKING AROUND THE WORLD

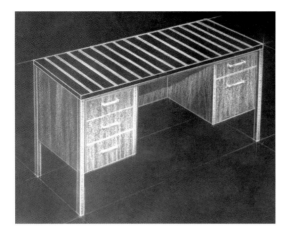

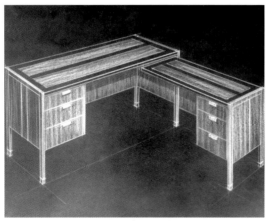
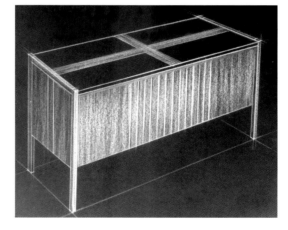

THIS PAGE: Renderings of small office desk proposals (1960s)

NEXT PAGE: Renderings of a convertible chair-bed concept (1960s)

WORKING AROUND THE WORLD

A secretarial chair with arms for Sears Office Supply Division. The division represented a large business segment for the company because small home-office owners could use their charge cards to purchase equipment. Sears charge card business became a large revenue source for the company. The chair, manufactured in Japan, was made of chrome-plated steel for sparkle and covered with bright nylon fabric. (1964)

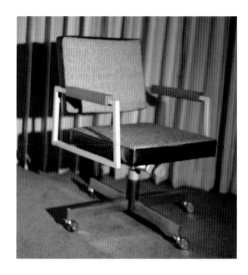

Executive chair for Sears made by H.O.N Company (Home-O-Nise) in Muscatine, Iowa. H.O.N. later grew to become an industry leader because of Stanley Howe, CEO (then president) and Phil Temple, chief engineer. They were given a great boost by Jack Milne, a senior buyer who was probably the best all-around merchant at Sears or in the country, but he was not political enough to become president of the company. (1965)

WORKING AROUND THE WORLD

This portable electric-powered shoe polisher was made by Iona Manufacturing Company for the Sears product line. (1972)

This guitar case, designed for maximum protection of the instrument, used the blow molding process, a method that became more common in the late 1960s. The top half lifted off, instead of tipping on hinges like most cases. (1968)

Console hair dryer for counter or tabletops. Loaded with features including three adjustable mirrors, a built-in hose with bonnet, and a light dimmer. Housed in a teak cabinet with brushed aluminum handle and trim. (1978)

WORKING AROUND THE WORLD

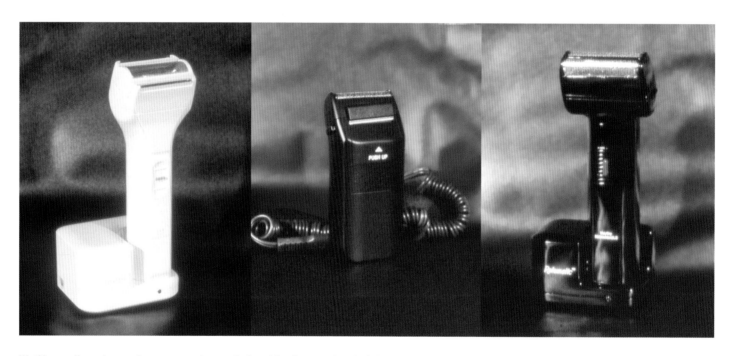

Wet/dry cordless shavers for women and men; designed for Sears and made in Hong Kong. (1991)

 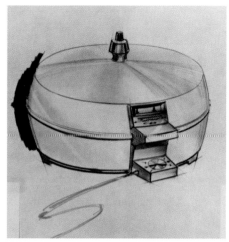 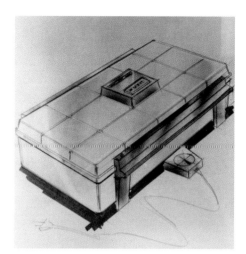

ABOVE: Concept sketches for tabletop cooking products for Sears. (1970)

These slimline toasters for Sears held two pieces of bread end-to-end rather than side-by-side. These toasters were part of the Kenmore line, an exclusive Sears brand. (1971)

WORKING AROUND THE WORLD

This oral hygiene unit was designed for the Sears' New York office and developed, in part, with General Electric (Bridgeport, Connecticut) and their Design and Engineering Department (North Carolina). This would have been the first Sears/GE partnership, if not for management loggerheads. The renderings to the right were concepts for water-driven oral hygiene units that included a cordless toothbrush. (1972)

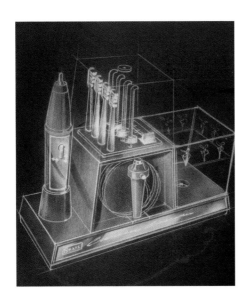

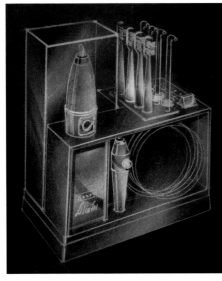

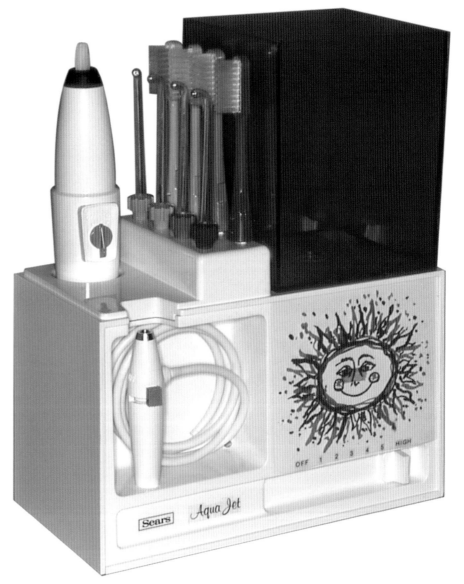

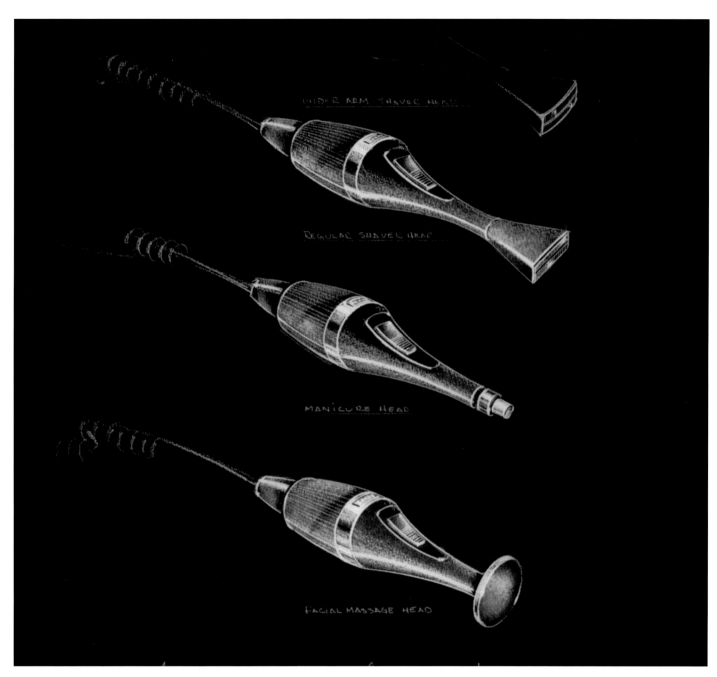

Renderings of proposals for manicure products for Sears personal care unit (1966)

WORKING AROUND THE WORLD

A very compact portable hair dryer for travel. The hose leading to the flexible bonnet stores inside the unit, and the removable power cord and bonnet are stored in a carrying case. (1978)

Low-end electric hair curler made of powder coated steel and plastic. Made in Hong Kong for Sears. (1987)

Deluxe electric hair curler made of styrene plastic and chrome-plated steel. Made in Hong Kong by Semitelex Manufacturing Company. (1988)

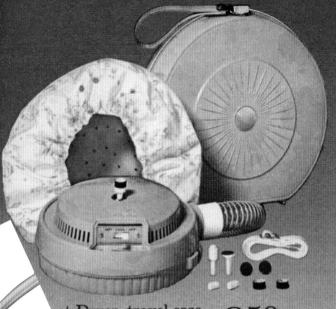

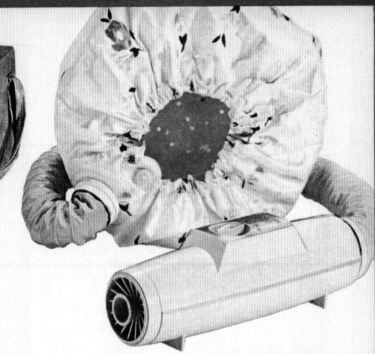

...eat Dryer, travel case ...3-piece manicure kit $9.50

Handy walk 'n wear style with carrying strap, 7-foot cord to let you move around while hair dries. Dial cool or hot. Fast-drying Beehive Bonnet ends irritating hot spots .. dries evenly.

For a speedy manicure .. insert attachments into power center of dryer. Fine file, nail shaper and buffer attachments are included.

Rigid plastic hatbox, mirror in lid. 225 watts.
8 K 8791—Shipping weight 6 lbs........ $9.50

1½ pounds light .. yet choice of 4 heats plus 500 watts of heating power assure fast, comfortable drying

Jiffy Jet $19.50

Slim .. trim .. lightweight .. yet powerful. Our tiny portable is just 10½ inches long, 2½ inches in diameter. Carrying strap plus 12-foot cord give you real walk 'n wear convenience.

To touch up droopy ends without disturbing entire hairdo, just attach Quick-Curl cylinder to hose. 3 plastic rollers included.

Beehive Bonnet speeds drying 35% over side-lined hoods we tested. To speed drying further, collapsible vinyl hose has wider diameter to force more air into bonnet .. stores in dryer.

Handsome travel case has adjustable carrying strap .. fold-over storage compartment. Leather-look vinyl over chipboard.
8 K 8745—Shipping weight 4 pounds................ $19.50

NOTE: Dryers UL listed; 110-120 volt, 60-cycle AC.

Quick-Curl Attachment for touch-up drying

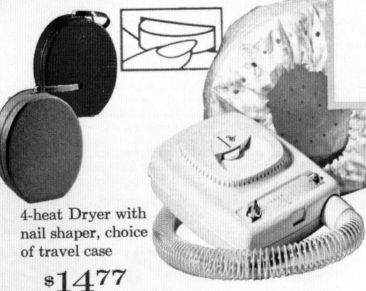

4-heat Dryer with nail shaper, choice of travel case $14.77

Just 2¼ pounds light with handy shoulder strap plus 8-foot cord for walk 'n wear convenience. Emery wheel spins to expertly shape your nails .. special vent dries nail polish. Our exclusive Beehive Bonnet dries 35% faster than side-lined hoods we tested. Uses 275 watts. Zipper hatbox of leather-look vinyl over chipboard .. pocket in lid for bonnet.
8 K 8788—Dryer with gold-color-flecked blue case. Shpg. wt. 5 lbs. 4 oz....... $14.77
8 K 8798—Dryer with red alligator-look case. Shpg. wt. 5 lbs. 4 oz............ 14.77

Walk 'n Wear 2-heat Dryer $6.98

Dial cool or hot. Carrying strap plus 7-foot cord let you cook, clean or iron while hair dries.

Our pretty Beehive Bonnet dries 35% faster than side-lined hoods we tested .. fits over large rollers.

Flexible plastic hose locks onto hood, dryer. Uses 225 watts.
Shpg. wt. 5 lbs.
8 K 8796....... $6.98

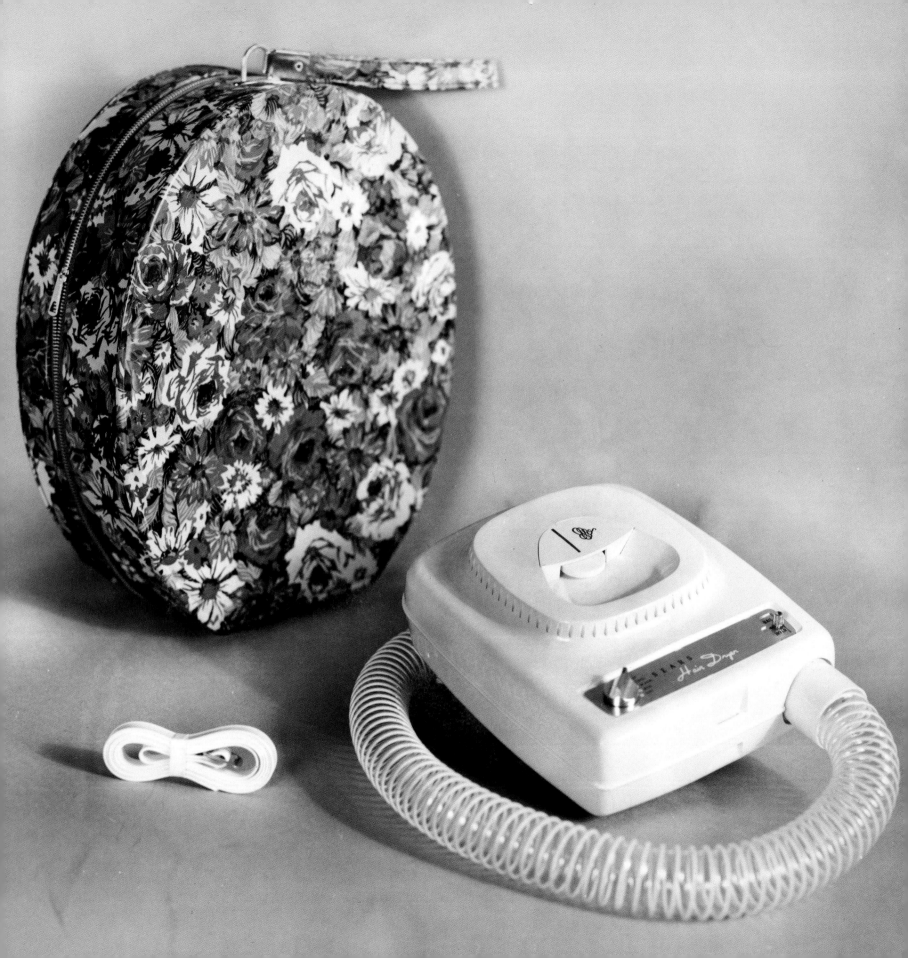

OPPOSITE: This portable hair dryer went into production and was a sales success. Features included an emery wheel built into the surface for manicures.

THIS PAGE: Tabletop portable hair dryers.

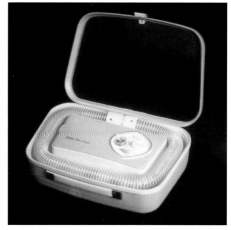
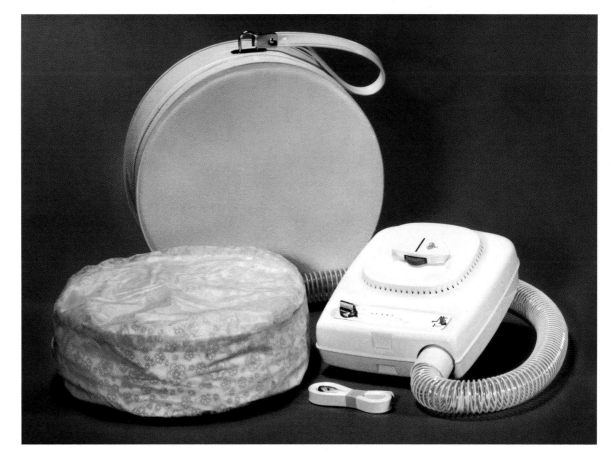

WORKING AROUND THE WORLD

THIS PAGE AND NEXT: When Sears needed a new exclusive look in a hair dryer with distinctive features, this unit was the result. The dryers featured a built-in power manicure set, remote control for heat setting, built-in cord storage and a smoke-tinted skirt on the bonnet to give a more comfortable feeling for the user. (1977)

WORKING AROUND THE WORLD

This portable whirlpool bath appliances was an inexpensive solution for people who wanted the benefits of a whirlpool bath for therapy or pleasure. The motor hung outside the tub and the impeller, on a stem inside the tub, drove water. Manufactured by Dazey Products of Kansas City, Missouri. (1975)

This heated foot bath was made by Dazey Products, a Kansas City manufacturer. Henry Talge, owner of Dazey Products, was a character. He had President Harry Truman's chair in his office and he yelled, "Don't sit in that chair! It's Harry's chair." Henry drove a gold-plated Cadillac. He said that the only other one was driven by Father Divine. Henry used to send me invitations to a birthday party in honor of Harry Truman for years and years even after Truman's death. (1975)

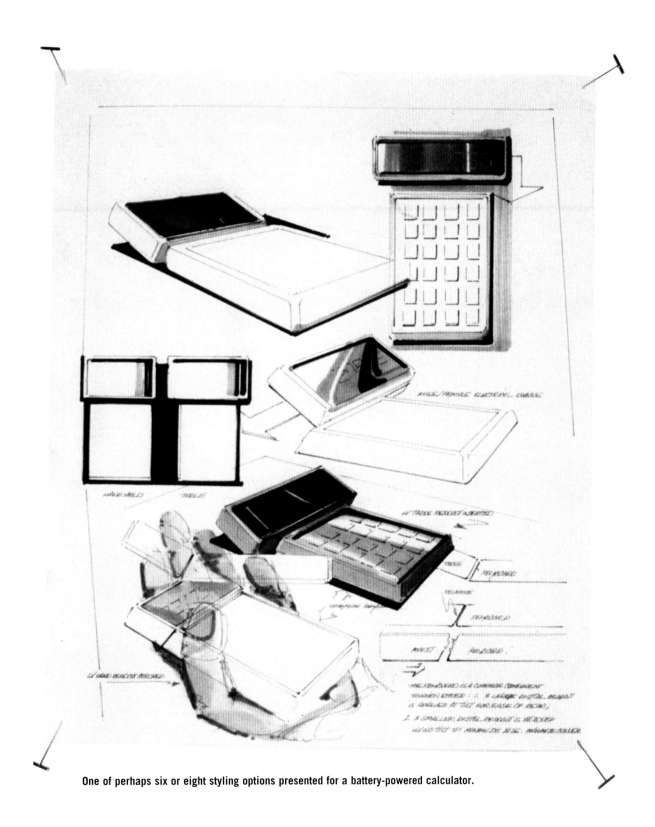

One of perhaps six or eight styling options presented for a battery-powered calculator.

WORKING AROUND THE WORLD

THIS PAGE AND NEXT: These bandsaw and radial saw design proposals updated to assure continuity with the company's marketing direction. The concepts were developed under my direction. (1991)

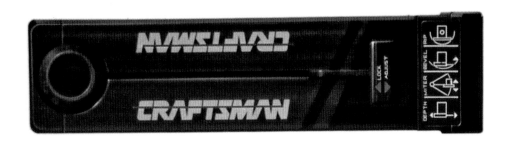
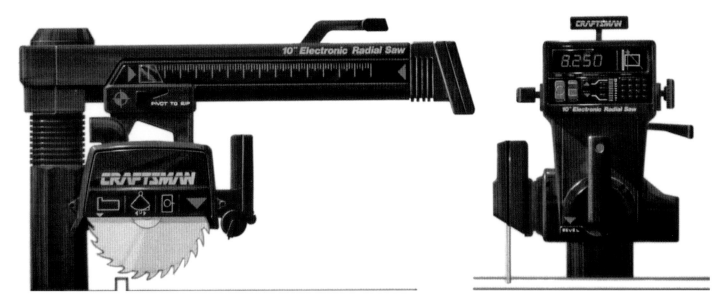

WORKING AROUND THE WORLD

This lawn and garden cart followed the success of the Sears Plastic Refuse Container. It shared many of the same features: blow-molded wheels, nested for shipping, did not rust or dent, and the handle stored inside when not in use. (1970)

This gas-powered leaf and grass blower, developed for Sears in conjunction with Baird-Poulan, was made with a die cast aluminum housing and blow-molded impeller cover and spout. (1990)

ABOVE AND RIGHT: This electric-powered leaf and grass blower and electric hedge trimmer were made of blow-molded and injection-molded parts in order to maximize economical manufacturing and maintain light weight. (1991)

WORKING AROUND THE WORLD

Gas-powered shredder for garden use. (1980)

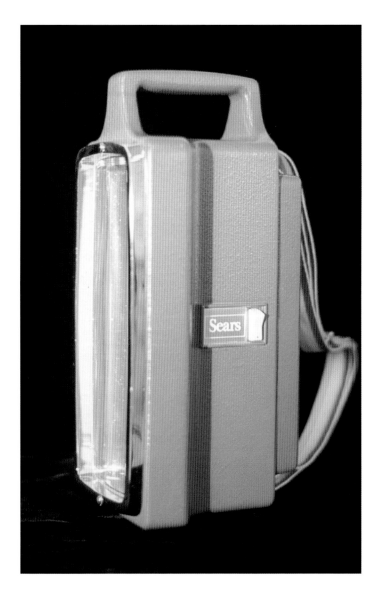 

This battery-powered lantern was redesigned during a time when many camping products were changing material from metal to plastic for weight and cost advantages. This lantern was made of blow-molded polypropylene. (1987)

WORKING AROUND THE WORLD

OPPOSITE-TOP: Rear discharge, gas-powered, self-propelled lawn mower developed in conjunction with a consultant.

OPPOSITE-MIDDLE: Walk-behind, electric, side discharge mower.

OPPOSITE-BOTTOM: Early electric mower designed for Simpson/Sears of Canada. This mower has a flip over handle to help keep the power cord on the same side of the mower while in use.

TOP: Rear discharge, electric, self-propelled lawn mower.

BELOW: Walk-behind, self-propelled, side discharge mower designed to integrate the gas tank with the other engine components for a more unified appearance.

BOTTOM: Electric mower with mulching deck.

WORKING AROUND THE WORLD

THIS PAGE AND NEXT: This garden tiller and lawn tractor were made for Sears by American Yard Products (AYP) in Orangeburg, South Carolina. They were designed in collaboration with the AYP design department headed by Brian Coleman.

"Chuck's not likely to blow his own horn or represent himself as being more than he is. His description of himself would have to be enhanced by someone else to be completely accurate. He's not only not going to take credit for things he didn't do, but he's not going to take credit for all the things he did do. I mean, he's going to tell you the things that are important, but not for the purpose of embellishing them in order to give an impression."

~Bob Johnson was a successful Sears & Roebuck senior vice president. He co-founded Johnson Bryce Corporation in 1991.

In 1992, Sears moved from the Sears Tower to Hoffman Estates, a northwest suburb of Chicago. That same year Janet became ill. She was diagnosed with multiple myeloma, a type of cancer. I wanted to retire to be more available for Janet. Sears was still downsizing and offering employees retirement incentive packages to leave the company. However, I was not able to get one, because management told me that if I left, it would have to end the design function within the corporation. At the time, there were only four of us in the design department.

By January of 1993, I was the only member of the design department. While Janet and I were in Hawaii on vacation, the corporation announced a broad, sweeping retirement incentive offer that was not selective. The people on the west coast of the mainland must have been able to hear me shout for joy when I heard the news.

I retired that year, and went home to take care of Janet. A lot of other people took the package, too—resulting in the largest exodus Sears ever had.

I spent most of my time adjusting to retirement and providing as much comfort to Janet as I could. She was having a tough time. Before I retired, our son, Charley, had moved his office into our basement so he could be with her during the day. I stayed with her at night. She was consistently getting radiation and chemotherapy treatments, but by June, she was showing signs of improvement.

Later that year, I received a call from the University of Illinois at Chicago (UIC), asking if I would consider teaching part-time. I had taught there on a part-time basis two or three years earlier while I was still at Sears. Janet's cancer was in remission by that time, so she could stay alone for lengthy periods. Also, I really needed a break from the house. The school asked me to take a full-time teaching load in the industrial design department, but I decided to teach part-time and was assigned the class that prepared senior students for entry into the industrial design profession.

Life was pretty much consumed with Janet's treatments. She went in and out of remission and it seemed to me that she was undergoing constant treatment or surgery. Around 1998, stem cell transplant seemed to be a possible option to help her. Although she was at the high end of the age scale for the procedure, her doctors thought it was possible and proceeded to harvest her stem cells in preparation for the treatment. They needed to gather enough cells to be able to perform two or three transplants. However, because of the chemotherapy treatments, her body was only able to produce enough cells for one implant treatment. So, the medical group decided not to proceed.

In late 1998, the myeloma returned aggressively. I was devastated when Janet passed away in November of 1999. We had been married for 42 years, and she was the driving force in my life, both professionally and personally. She was my reason for living. She was my best friend. She showed me the good things in life. She kept me balanced during difficult times, and she was always there for me. She was my sounding board and advisor for everything I faced. It seemed that I'd lost any reason for continuing with our short- and long-term plans, and my life had been turned upside down.

THE NEXT STEP

Within a matter of months after Janet passed, our son and his wife relocated to southern California in order to expand his career as a composer and find work scoring films. I joined a grief support group, which helped me considerably in learning how to work through grieving and how to go on with life.

Since teaching was one constant that provided some structured activity, a close friend and fellow professor at UIC, Twiley Barker, advised me to continue teaching at least for the short term. A sense of order slowly returned to my daily routine. I started an exercise regimen and got more involved in volunteer activities in my community and with Sears retirement groups.

I was still at Sears when I first became involved with the Over the Rainbow Association, a not-for-profit organization that helps people with physical disabilities help themselves. It asked Sears for some technical support to help put together a 33-unit apartment complex in Evanston, Illinois and I was called in as a designer. There were also a couple of engineers from different disciplines, and we worked on kitchens and lifts.

We teamed with Whirlpool and other Sears suppliers to modify kitchen appliance products. For instance, we'd put in refrigerator shelves that would pull out like file drawers for people in wheelchairs who hadn't been able to reach into the refrigerator, or we'd raise the dishwashers from the floor to wheelchair height. We worked on kitchen sinks so you could get a wheelchair under there. We had cabinets that would lower so that people could reach into them for food or dishes. We created lifts so people could be lifted from the bed to a wheelchair or from a wheelchair to a toilet or shower without the need for a strong assistant.

I stayed on for a long time as an advisor, but then I put my energies into a different problem. The chairman of Sears, Arthur Martinez, and his legal beagles decided to eliminate life insurance benefits for retirees. He reduced every retiree's benefits to $5,000. That got me riled up.

The average for Sears retirees had been between $17,000 and $20,000—many of us had more. What really got me was that some of those older people who had retired didn't have anything else except Social Security. I knew a few, in fact, who had gone into nursing homes and who had posted their insurance plan as collateral, so that when they died the nursing home would be able to recover the money. But Martinez came along and took it away.

So I was ready to go to war—whatever it took to attract attention. I joined a new group at Sears, NARSE (the National Association of Retired Sears Employees), and I'm still on the board. I marched up and down, carried signs and demonstrated against Sears management's decision. Because Sears was a significant financial supporter of Over the Rainbow, I thought I would put Rainbow in jeopardy, so I retired from its board.

With NARSE, we were on television and public radio. I was in Atlanta carrying picket signs—everywhere Sears had its board of directors' meeting. We wore yellow T-shirts that read, "Sears Unfair to Employees." We marched at shopping centers and showed up at airports where these guys were coming in. We were really an aggravation, but to no avail. Martinez never changed his decision. And the courts supported him. That was a shock to me. I was just as naïve as most Americans in thinking the justice system was equal and fair, but I learned that it could be bent just like anything else. Anyway, I still do things with Over the Rainbow when they call.

Also, I pursued lessons to learn Spanish, which had long interested me, but was difficult because of my dyslexia.

A year or so after Janet passed away, I sold our long-time home and moved into a small condominium, continuing to teach until early 2002, when I took a leave of absence from UIC. I wanted to travel a bit. Later that year, Columbia College in Chicago called and asked me to teach there, which I'm still doing. I'm teaching a senior class, Professional Practices for Product Design, which introduces students to the real world.

I've been president of the Chicago chapter of IDSA (Industrial Designers Society of America) and am a senior advisor for the Organization of Black Designers (OBD), which is based in Washington, D.C. I was involved in forming OBD's Chicago chapter in 1994 and really admire this group, which is very professional, very sharp. It supports the Chicago public schools with various mentoring programs in art and design, and also sponsors scholarships and workshops. Despite its name, its members include Caucasians, Asians and Native Americans.

The summer of 2003 was a sad one for me because my teacher, mentor and good friend, Henry Glass, passed away due to heart failure at the age of 92. I told the obituary writer for the *Chicago Tribune:* "He impressed on me that I was undertaking a serious career that could affect the lives of thousands of people. He was a tough taskmaster, but he encouraged people when he recognized success."

In the fall of 2003, I was one of the honorees—and sat at the Sears table—at the Executive Leadership Foundation's 15th Annual Recognition Dinner in Washington, D.C. The theme of the fundraising awards gala, "Celebrating Black Corporate Leadership, Heroes Past & Present," highlighted the mission and legacy of the Executive Leadership Council (ELC) and the pioneering achievements of African-American corporate executives from the 1940s through the present. Founded in 1986, the ELC is the country's premier organization of the most senior African-American corporate executives in Fortune 500 companies.

As for my non-professional interests, I've always been fascinated by water. In Texas, there was a creek nearby and I made a boat out of wood. Many years later, around 1969, when I was working at Sears, where they designed and sold boats, I began to tweak up an interest because a couple of the other designers were sailors. Initially, I bought a small boat that I could carry on top of my car. I often took it out to the forest preserves around Chicago and sailed around for a couple of hours. My family wasn't

Sailing crew and partners at a celebration honoring Edgar Hawley, senior partner on the 30-ft. sloop "The Moonlight".

GIVING OF MY GIFT

The "Next Step", a very fast 32-foot sloop sailboat built on a 5.5m racing hull. A boat co-owned with friends Ed Hawley and Bob Johnson. (1975)

really interested, but they tolerated me. Eventually, I bought an interest in a 24-foot trimaran (three hulls), and I've probably owned five or six boats since then. Now, I own a 22-footer that I keep in Wilmette Harbor, just north of my home.

I used to go sailing with Carl Boelke, a Sears colleague who calls me a "cautious sailor who knows the rules of the road"—which I can't say the same about other people I sailed with!" These days, I sail probably two or three days a week on Lake Michigan during the warm months. I've sailed to ports like Saugatuck and South Haven (both in Michigan); Michigan City, Indiana; and Milwaukee, Wisconsin. It's been a good thing for me. Now, most of the time I sail alone—at night with the stars, quiet. I just like being in the water and the wind and letting nature take me. You're never in control, really. It's a compromise which I really kind of like.

DESIGN AS A PROFESSION

Today, the design profession is so international that I don't think a single person or country influences it anymore. However, I do believe there's still a distinction in product development. Three major forces in the world can be identified very clearly.

First, the European market drives product that is based on technology. But it gets so involved in technology, that sometimes the price becomes prohibitive. Also, it gets so committed to advancing the technology, that the product also becomes impractical—not only from the user's standpoint, but more specifically, from the *repair* and *maintenance* standpoints.

Second, in Japan, Taiwan, South Korea and China, the focus really is on quality. For years after World War II, their products were considered inferior in the international marketplace, but they successfully built up the quality image, more so than most of the world. As a by-product of that, they also took an interest in the consumer and how the consumer interacted with their products; making them user-friendly. So not only do they maintain a quality standard as good as everybody else's, they also have the customer interaction standard that's higher. That's why I still buy Japanese automobiles.

Third, the American market is driven by cost. We introduce things to make them inexpensive, to sell as many as we can make. This is not meant to be negative. What it does for the American people is create products inexpensive enough for everyone to have something. Many European and Asian countries make a few products, and the cost is so high that everyone can't afford them. In the United States, you can buy a toaster for $10, but you can also buy one for $200. So our population can enjoy a good life, and I think that's a big plus.

TODAYS DESIGN CURRICULUM

If I were to change the curriculum for today's design students, I'd point out some fundamental things that I totally disagree with. Primarily, it's that because of computers, educators don't teach kids to draw. The thinking process from the inception of an idea through to a finished product uses a communication trail that is sacrificed because the computer software limits the extent of what you can do. So the moment the designer relinquishes his capability to software, he's limited the possibilities that are available. I can almost look at a product or even a graphics piece and tell you if it was done by hand or by computer. I feel an absence of *fluidity,* which is the closest word I can find for it.

I want to identify with things in nature. Those are the beautiful parts of life—plants, stars, ripples, sky, cloud formations, a blade of grass, and waves in the water. I don't see much reflection of that influence in product design these days.

I would also address developing marketing and business skills, recognizing that consumers have better taste than people give them credit for. Corporate executives, with some exceptions, don't have any training in the arts. I've been in circumstances where I would take in a design proposal, and the executive would call his secretary and ask her for her opinion. Or he'd take it home to his wife to see what she thought. They are not only off base there, but also in hundreds of other decisions they have to make. They're far out of touch, because that's the way the game is played. And it's not just in the U.S. It's universal.

MY PHILOSOPHY OF DESIGN AND LIFE

If there is anything distinctive or unique about my design philosophy or approach, I don't think it is clearly evident when looking at the finished products I designed. I can sum it up, however, with the following principles:

Esthetics—I strongly believe that things should not be forced to look a certain way. Appearance should evolve or come about naturally as a result of function. The visual statement should express a harmony with why the product exists—what it does, how it is made, what it is made of—and look pleasing if not beautiful.

Ingenuity—I believe that design should search all possible ways to maximize the usefulness of the result. This built-in versatility means that it may provide more than one function. If it is a product, the materials and manufacturing process used must be creatively explored and mated in a way that incorporates all opportunities.

Sincerity and Honesty—These design principles speak to lasting value. The honest designer must be conscious of not introducing fake or insincere elements—imitations of natural materials or components such as knobs or handles that don't work—that may add visual interest but have no connection to function. For me, this disconnect defined the difference between simple cosmetics and genuine esthetics.

And when one designs a product, the effort must enhance the quality of peoples' lives. In other words—and I firmly believe in this—contribute what you can, excluding trickery, cuteness, imitation or misrepresentation.

The intention is to strive for a straightforward, clear and simple solution. When intent is adhered to throughout the design process, the end result will project the image that the product "belongs" and is there to help and contribute.

When considering my life's design, I have to say that as with the works that I have tried to create, shape or improve, form has arisen from use. And use, or

GIVING OF MY GIFT

usefulness, has gone beyond mere circumstance and has had its root buried deep in the soil of need. Circumstance dictated that a boy from the rural South, a black boy whose family lived on the fine edge between poverty and doing all right, was destined to neither amount to nor contribute much. Add to that a condition that forced me to train my mind to literally turn objects around so that I could discover their meaning.

But I came from a family that taught me to see beyond circumstance, find value and create usefulness out of need. The legacy I inherited required that I make myself useful—as a son, man, husband, father and artist.

If I were to share one thought with the design community of today and tomorrow it would be to remember that your purpose—your gift to the world—is to provide straightforward solutions to real problems for living, breathing human beings. As an industrial designer especially, your audience is neither history nor fame, but a couple who worked hard to buy their first home on a quiet street and would love just one more hour of sleep in the morning, even on trash day. Your muse is the kid who needs something to occupy his mind and hands during that long drive to grandma's house. Your biggest critic will be the struggling mother who can't afford to keep replacing her kitchen appliances every time a little piece of ornate but useless piece of plastic breaks off.

Know the difference between deep satisfaction and simply delivering a 'wow.' The latter is fancy and derives from simply delivering the unexpected. On the other hand, I believe that deep satisfaction arises when you find an elegant solution to a problem that has, until now, had a hindering effect or negative impact on a person's quality of life or experience. The elegant solutions I am talking about should be executed in such a straightforward manner as to nearly scream their presence to the world. Function should be obvious—a straightforward solution to a meaningful problem.

If I humbly submit my life's design as an example of anything, I would dare say that I have shown, through conscious effort and accident that human beings are the most creative when we encounter the unexpected. Software will give you exactly what you ask for. Corporations drive toward the bottom line, and sometimes the lowest common denominator. On the other hand, a pencil and paper might throw you a curve. A lack of job security may cause you to find security deep within yourself. The afternoon sun may literally shed light on a problem in a way that you may not have seen at daybreak. Being the odd man out may give you the right perspective. Learning to fish without a pole might just change your life.

745.2 HARRISON
Harrison, Charles,
A life's design : the
 life and work of industr

GENERAL COLLECTIONS

3RD FLOOR
CENTRAL LIBRARY

Atlanta-Fulton Public Library